BROOKLANDS BOOKS

MINI
MUSCLE CARS
1961-1979

Compiled by
R.M. Clarke

ISBN 0 907 073 697

Distributed by
Brooklands Book Distribution Ltd.
'Holmerise', Seven Hills Road,
Cobham, Surrey, England
Printed in Hong Kong

BROOKLANDS BOOKS

BROOKLANDS BOOKS SERIES
AC Ace & Aceca 1953-1983
AC Cobra 1962-1969
Alfa Romeo Alfasud 1972-1984
Alfa Romeo Alfetta Coupes GT.GTV.GTV6 1974-1987
Alfa Romeo Guilias Berlinettas
Alfa Romeo Giulia Berlinas 1962-1976
Alfa Romeo Giulia Coupés 1963-1976
Alfa Romeo Spider 1966-1987
Allard Gold Portfolio 1937-1958
Aston Martin Gold Portfolio 1972-1985
Austin Seven 1922-1982
Austin A30 & A35 1951-1962
Austin Healey 100 1952-1959
Austin Healey 3000 1959-1967
Austin Healey 100 & 3000 Collection No. 1
Austin Healey 'Frogeye' Sprite Collection No. 1
Austin Healey Sprite 1958-1971
Avanti 1962-1983
BMW Six Cylinder Coupés 1969-1975
BMW 1600 Collection No. 1
BMW 2002 1968-1976
Bristol Cars Gold Portfolio 1946-1985
Buick Automobiles 1947-1960
Buick Riviera 1963-1978
Cadillac Automobiles 1949-1959
Cadillac Automobiles 1960-1969
Cadillac Eldorado 1967-1978
Camaro 1966-1970
Chevrolet Camaro & Z-28 1973-1981
High Performance Camaros 1982-1988
Chevrolet Camaro Collection No. 1
Chevrolet 1955-1957
Chevrolet Impala & SS 1958-1971
Chevelle & SS 1964-1972
Chevy II Nova & SS 1962-1973
Chrysler 300 1955-1970
Citroen Traction Avant 1934-1957
Citroen DS & ID 1955-1875
Citroen 2CV 1948-1988
Cobras & Replicas 1962-1983
Cortina 1600E & GT 1967-1970
Corvair 1959-1968
Daimler Dart & V-8 250 1959-1969
Datsun 240z 1970-1973
Datsun 280Z & ZX 1975-1983
De Tomaso Collection No. 1
Dodge Charger 1966-1974
Excalibur Collection No. 1
Ferrari Cars 1946-1956
Ferrari Cars 1962-1966
Ferrari Cars 1969-1973
Ferrari Dino 1965-1974
Ferrari Dino 308 1974-1979
Ferrari 308 & Mondial 1980-1984
Ferrari Collection No. 1
Fiat-Bertone X1/9 1973-1988
Fiat Pininfarina 124+2000 Spider 1968-1985
Ford Falcon 1960-1970
Ford Mustang 1964-1967
Ford Mustang 1967-1973
High Performance Mustangs 1982-1988
Ford RS Escort 1968-1980
Honda CRX 1983-1987
High Performance Escorts MkI 1968-1974
High Performance Escorts MkII 1975-1980
Hudson & Railton Cars 1936-1940
Jaguar XK120 XK140 XK150 Gold Portfolio 1948-1960
Jaguar Cars 1957-1961
Jaguar Cars 1961-1964
Jaguar MK2 1959-1969
Jaguar E-Type 1961-1966
Jaguar E-Type 1966-1971
Jaguar E-Type V12 1971-1975
Jaguar XKE Collection No. 1
Jaguar XJ6 1968-1972
Jaguar XJ6 Series II 1973-1979
Jaguar XJ6 & XJ12 Series III 1979-1985
Jaguar XJ12 1972-1980
Jaguar XJS Gold Portfolio 1975-1988
Jensen Cars 1946-1967
Jensen Cars 1967-1979
Jensen Interceptor Gold Portfolio 1966-1986
Lamborghini Cars 1964-1970
Lamborghini Cars 1970-1975
Lamborghini Countach Collection No. 1
Lamborghini Countach & Urraco 1974-1980
Lamborghini Countach & Jalpa 1980-1985
Lancia Stratos 1972-1985
Land Rover 1948-1973
Land Rover Series II & IIa 1958-1971
Land Rover Series III 1971-1985
Land Rover 90 & 110 1983-1989
Lotus Cortina 1963-1970
Lotus Elan Gold Portfolio 1962-1974
Lotus Elan Collection No. 2
Lotus Elite 1957-1964
Lotus Elite & Eclat 1974-1981
Lotus Turbo Esprit 1980-1986
Lotus Europa 1966-1975
Lotus Europa Collection No. 1
Lotus Seven 1957-1980
Lotus Seven Collection No. 1
Maserati 1965-1970
Maserati 1970-1975
Marcos Cars 1960-1988
Mazda RX-7 Collection No. 1
Mercedes 190 & 300SL 1954-1963
Mercedes 230/250/280SL 1963-1971
Mercedes 350/450SL & SLC 1971-1980
Mercedes Benz Cars 1949-1954
Mercedes Benz Cars 1954-1957
Mercedes Benz Cars 1957-1961
Mercedes Benz Competition Cars 1950-1957

Metropolitan 1954-1962
MG Cars 1929-1934
MG TC 1945-1949
MG TD 1949-1953
MG TF 1953-1955
MG Cars 1957-1959
MG Cars 1959-1962
MG Midget 1961-1980
MGA Collection No. 1
MGA Roadsters 1955-1962
MGB Roadsters 1962-1980
MGB GT 1965-1980
Mini Cooper 1961-1971
Morgan Cars 1960-1970
Morgan Cars 1969-1979
Morris Minor Collection No. 1
Old's Cutlass & 4-4-2 1964-1972
Oldsmobile Toronado 1966-1978
Opel GT 1968-1973
Packard Gold Portfolio 1946-1958
Pantera 1970-1973
Pantera & Mangusta 1969-1974
Plymouth Barracuda 1964-1974
Pontiac Fiero 1984-1988
Pontiac GTO 1964-1970
Pontiac Firebird 1967-1973
Pontiac Firebird and Trans-Am 1973-1981
High Performance Firebirds 1982-1988
Pontiac Tempest & GTO 1961-1965
Porsche Cars 1960-1964
Porsche Cars 1964-1968
Porsche Cars 1968-1972
Porsche Cars in the Sixties
Porsche Cars 1972-1975
Porsche 356 1952-1965
Porsche 911 Collection No. 1
Porsche 911 Collection No. 2
Porsche 911 1965-1969
Porsche 911 1970-1972
Porsche 911 1973-1977
Porsche 911 Carrera 1973-1977
Porsche 911 SC 1978-1983
Porsche 911 Turbo 1975-1984
Porsche 914 Gold Portfolio 1969-1988
Porsche 914 Collection No. 1
Porsche 924 1975-1981
Porsche 928 Collection No. 1
Porsche 944 1981-1985
Porsche Turbo Collection No. 1
Reliant Scimitar 1964-1986
Riley 1½ & 2½ Litre Gold Portfolio 1945-1955
Rolls Royce Silver Cloud 1955-1965
Rolls Royce Silver Shadow 1965-1980
Range Rover Gold Portfolio 1970-1988
Rover P4 1949-1959
Rover P4 1955-1964
Rover 2000 + 2200 1963-1977
Rover 3500 1968-1977
Rover 3500 & Vitesse 1976-1986
Saab Sonett Collection No. 1
Saab Turbo 1976-1983
Studebaker Hawks & Larks 1956-1963
Sunbeam Tiger And Alpine Gold Portfolio 1959-1967
Thunderbird 1955-1957
Thunderbird 1958-1963
Thunderbird 1964-1976
Toyota MR2 1984-1988
Triumph 2000-2.5-2500 1963-1977
Triumph Spitfire 1962-1980
Triumph Spitfire Collection No. 1
Triumph Stag 1970-1980
Triumph Stag Collection No. 1
Triumph TR2 & TR3 1952-1960
Triumph TR4.TR5.TR250 1961-1968
Triumph TR6 1969-1976
Triumph TR6 Collection No. 1
Triumph TR7 & TR8 1975-1982
Triumph GT6 1966-1974
Triumph Vitesse & Herald 1959-1971
TVR Gold Portfolio 1959-1988
Volkswagen Cars 1936-1956
VW Beetle 1956-1977
VW Beetle Collection No. 1
VW Golf GTi 1976-1986
VW Karmann Ghia 1955-1982
VW Scirocco 1974-1981
VW Bus-Camper-Van 1954-1967
VW Bus-Camper-Van 1968-1979
Volvo 1800 1960-1973
Volvo 120 Series 1956-1970

BROOKLANDS MUSCLE CARS SERIES
American Motors Muscle Cars 1966-1970
Buick Muscle Cars 1965-1970
Camaro Muscle Cars 1966-1972
Capri Muscle Cars 1969-1983
Chevrolet Muscle Cars 1966-1972
Dodge Muscle Cars 1967-1970
Mercury Muscle Cars 1966-1971
Mini Muscle Cars 1961-1979
Mopar Muscle Cars 1964-1967
Mopar Muscle Cars 1968-1971
Mustang Muscle Cars 1967-1971
Shelby Mustang Muscle Cars 1965-1970
Oldsmobile Muscle Cars 1964-1970
Plymouth Muscle Cars 1966-1971
Pontiac Muscle Cars 1966-1972
Muscle Cars Compared 1966-1971
Muscle Cars Compared Book 2 1965-1971

BROOKLANDS ROAD & TRACK SERIES
Road & Track on Alfa Romeo 1949-1963
Road & Track on Alfa Romeo 1964-1970
Road & Track on Alfa Romeo 1971-1976

Road & Track on Alfa Romeo 1977-1984
Road & Track on Aston Martin 1962-1984
Road & Track on Auburn Cord & Duesenberg 1952-1984
Road & Track on Audi 1952-1980
Road & Track on Audi 1980-1986
Road & Track on Austin Healey 1953-1970
Road & Track on BMW Cars 1966-1974
Road & Track on BMW Cars 1975-1978
Road & Track on BMW Cars 1979-1983
Road & Track on Cobra, Shelby & Ford GT40 1962-1983
Road & Track on Corvette 1953-1967
Road & Track on Corvette 1968-1982
Road & Track on Corvette 1982-1986
Road & Track on Datsun Z 1970-1983
Road & Track on Ferrari 1950-1968
Road & Track on Ferrari 1968-1974
Road & Track on Ferrari 1975-1981
Road & Track on Ferrari 1981-1984
Road & Track on Fiat Sports Cars 1968-1987
Road & Track on Jaguar 1950-1960
Road & Track on Jaguar 1961-1968
Road & Track on Jaguar 1968-1974
Road & Track on Jaguar 1974-1982
Road & Track on Jaguar 1983-1989
Road & Track on Lamborghini 1964-1985
Road & Track on Lotus 1972-1981
Road & Track on Maserati 1952-1974
Road & Track on Maserati 1975-1983
Road & Track on Mazda RX7 1978-1986
Road & Track on Mercedes 1952-1962
Road & Track on Mercedes 1963-1970
Road & Track on Mercedes 1971-1979
Road & Track on Mercedes 1980-1987
Road & Track on MG Sports Cars 1949-1961
Road & Track on MG Sports Cars 1962-1980
Road & Track on Mustang 1964-1977
Road & Track on Peugeot 1955-1986
Road & Track on Pontiac 1960-1983
Road & Track on Porsche 1951-1967
Road & Track on Porsche 1968-1971
Road & Track on Porsche 1972-1975
Road & Track on Porsche 1975-1978
Road & Track on Porsche 1979-1982
Road & Track on Porsche 1982-1985
Road & Track on Rolls Royce & Bentley 1950-1965
Road & Track on Rolls Royce & Bentley 1966-1984
Road & Track on Saab 1955-1985
Road & Track on Toyota Sports & GT Cars 1966-1986
Road & Track on Triumph Sports Cars 1953-1967
Road & Track on Triumph Sports Cars 1967-1974
Road & Track on Triumph Sports Cars 1974-1982
Road & Track on Volkswagen 1951-1968
Road & Track on Volkswagen 1968-1978
Road & Track on Volkswagen 1978-1985
Road & Track on Volvo 1957-1974
Road & Track on Volvo 1975-1985
Road & Track Henry Manney At Large & Abroad

BROOKLANDS CAR AND DRIVER SERIES
Car and Driver on BMW 1955-1977
Car and Driver on BMW 1977-1985
Car and Driver on Cobra, Shelby & Ford GT40 1963-1984
Car and Driver on Datsun Z 1600 & 2000 1966-1984
Car and Driver on Corvette 1956-1967
Car and Driver on Corvette 1968-1977
Car and Driver on Corvette 1978-1982
Car and Driver on Corvette 1983-1988
Car and Driver on Ferrari 1955-1962
Car and Driver on Ferrari 1963-1975
Car and Driver on Ferrari 1976-1983
Car and Driver on Mopar 1956-1967
Car and Driver on Mopar 1968-1975
Car and Driver on Mustang 1964-1972
Car and Driver on Pontiac 1961-1975
Car and Driver on Porsche 1955-1962
Car and Driver on Porsche 1963-1970
Car and Driver on Porsche 1970-1976
Car and Driver on Porsche 1977-1981
Car and Driver on Porsche 1982-1986
Car and Driver on Saab 1956-1985
Car and Driver on Volvo 1955-1986

BROOKLANDS MOTOR & THOROUGHBRED & CLASSIC CAR SERIES
Motor & T & CC on Ferrari 1966-1976
Motor & T & CC on Ferrari 1976-1984
Motor & T & CC on Lotus 1979-1983
Motor & T & CC on Morris Minor 1948-1983

BROOKLANDS PRACTICAL CLASSICS SERIES
Practical Classics on Austin A 40 Restoration
Practical Classics on Land Rover Restoration
Practical Classics on Metalworking in Restoration
Practical Classics on Midget/Sprite Restoration
Practical Classics on Mini Cooper Restoration
Practical Classics on MGB Restoration
Practical Classics on Morris Minor Restoration
Practical Classics on Triumph Herald/Vitesse
Practical Classics on Triumph Spitfire Restoration
Practical Classics on VW Beetle Restoration
Practical Classics on 1930S Car Restoration

BROOKLANDS MILITARY VEHICLES SERIES
Allied Military Vehicles Collection No. 1
Allied Military Vehicles Collection No. 2
Dodge Military Vehicles Collection No. 1
Military Jeeps 1941-1945
Off Road Jeeps 1944-1971
V W Kubelwagen 1940-1975

BROOKLANDS
BOOKS

CONTENTS

ACKNOWLEDGEMENTS

Some people have a gift for using one or two words in such a way as to immediately conjure up a mental picture of a new phenomenon. When coined these words cross all frontiers and are universally adopted, they are also improved upon when the need arises. An example of this is market which became supermarket with the introduction of larger self-service food stores and when the French improved on this concept they became hyper-markets.

On the automotive scene cars such as the Mustang and Camaro became known as 'Pony Cars' and when they were fitted with power units that would not disgrace a small tank they became 'Muscle Cars'.

The Issigonis Mini was introduced in 1959, a neat, small, economical bundle that was designed to transport Mr. & Mrs. Average with their 1.7 children around the British Isles. Because its road holding properties were exceptional it attracted the attention of that well known 'trainer' John Cooper who passed it through his gymn and the Mini-Cooper muscle car was shown to the world in September 1961. The stories in this book go on to tell just how he and others coaxed more power out of this tiny machine.

To illustrate the marque's prowess Mini-Coopers won the Monte Carlo Rally in 1964, 1965 and again in 1967. The car that pulled off this last victory is shown on the front cover, it was driven by Rauno Aaltonen and beat Vic Elford's Porsche 911 and Ove Andersson's Lancia Fulvia HF by 13 seconds. This car is now the property of BL Heritage and is frequently on show at their fine museum at Syon Park, Brentford, near London and we are indebted to them for allowing us to photograph it.

Brooklands Books are a historical reference series. They currently number some 150 titles and make available to today's drivers, owners and restorers a vast amount of information which in the main would be lost to enthusiasts. They could not appear without the understanding and goodwill of the publishers of the world's leading motor journals who for over twenty years have allowed the inclusion of their copyright articles in the series.

I am sure that Mini devotees will wish to join with me in thanking the management of Autocar, Autosport, Cars & Car Conversions, Motor, Motor Racing, Motor Sport, Motoring, Sports Car World, Thoroughbred & Classic Cars, Track & Traffic and Wheels for their continued support.

R.M. Clarke

A companion book Mini-Cooper 1961-1971 containing a completely different collection of articles is also available.

The Morris-Cooper (*left*) and Austin-Cooper(*right*) share the two-tone treatment of body and roof with the new Super models.

1962 Cars

B.M.C. COOPER MINIS

Disc - Braked 1,000c.c. Austin 7 Cooper and Morris Mini-Cooper. New Super de Luxe 850 Models

SINCE those identical twins the Austin 7 and Morris Mini-Minor were introduced in 1959, they have carved an extraordinary niche in the motoring world. Initial prejudice, an inevitable legacy from less practical miniature cars that had gone before, was soon dispelled when people found that the ADO 15 (its drawing office type number) with its unorthodox transverse engine, front wheel drive and square lines was only small outside. Inside it provides more rather than less room than the average light car for four people, with a performance well up to that of its competitors but with roadholding, stability and steering that completely transcend the performance. These last features have made it one of the most popular cars to be seen in current club racing and brought a new prosperity to specialist tuning establishments.

B.M.C. themselves have now stepped in to supply this growing demand for much more power with the new ADO 50 which will be called the Austin 7 Cooper or Morris Mini-Cooper. In developing this project they have had the assistance of the Cooper Car Co. who have amassed considerable experience of the A-type engines

fitted to their successful Formula Junior cars and developed to give some 85-90 b.h.p. Although such extreme tuning would be entirely inappropriate to a quiet, durable road car it very rapidly accelerates the process which Americans call "ruggedization." B.M.C. have been content with a 60% increase to 55 b.h.p. (net) at 6,000 r.p.m. which has been obtained by enlargement of the capacity from 848 c.c. to 997 c.c. and by alterations to the camshaft, carburation and cylinder head. A close-ratio gearbox, disc front brakes, special Dunlop tyres and better sound-proofing are other important modifications but the size, weight and appearance are almost unchanged.

Specially designed 7-in. Lockheed discs have $\frac{1}{4}$ in. thick pads each of $2\frac{3}{4}$ sq. in. area. The rubbed area is nearly twice as great as that of the standard 7 in. drums.

The increase of capacity comes from an entirely new crankshaft giving a stroke 6.51 mm. longer than that of the ADO 15. This shaft has thicker webs than the basic design, a larger diameter at the flywheel end and a torsional damper at the free end, an unusual fitting for a small 4-cylinder engine but a useful insurance against failure due to critical vibration periods which might be encountered if it is over-revved in the lower gears. The cylinder bore has actually been decreased by half a millimetre; the reason for this is not revealed but it keeps the swept volume inside the 1,000 c.c. mark at present whilst apparently giving the possibility of further enlargement to about 1,100 c.c. if this should be found necessary or desirable in the future.

Compression ratio has risen to 9.0 from 8.3 and the combustion chambers have been re-designed to improve gas flow; the new cylinder head is very similar to that of the Mk. II Austin-Healey Sprite. The exhaust valve head diameter remains the same as that of the 850 engine (1 in.) but the inlet valves have increased by 1/16 in. to $1\frac{5}{32}$ in. and both are now controlled by double valve springs. Two $1\frac{1}{4}$ in. S.U. carburetters supply mixture through enlarged ports and a free-flow 3-branch manifold with a large capacity silencer eases the exhaust gas flow. The new camshaft lengthens the total opening period of each valve from 230° to 252° with an overlap at T.D.C. of 37° instead of 15°. These changes have raised the maximum power from 34 at 5,500 r.p.m. to 55 b.h.p.

B.M.C. Cooper Minis

This interior view shows the new instrument panel, the remote control gear lever, re-designed seats and trim and the door-opening handles which have replaced wire pulls. The new Super 850-c.c. models are identical except for the gear lever.

at 6,000 r.p.m. with torque increases varying from 25% in the mid-speed range to 60% at 5,500 r.p.m. The highest b.m.e.p. is 135 lb. sq. in. at 3,600 r.p.m. but it remains above 118 lb. sq. in. all the way from 1,000 r.p.m. to a little over 6,000.

The final drive ratio remains the same in the ADO 50 but the much closer intermediate ratios of the new Sprite Mk. II are employed, giving reductions of 3.20, 1.918 and 1.375 in 1st, 2nd and 3rd gears. A remote-control gearchange linkage presents an almost vertical gear lever moving in a much more natural way for accurate control by the driver.

Special Gold Seal tyres have been developed by Dunlop to match the greater performance. The size (5.20-10) remains the same but a nylon casing provides the necessary strength for prolonged high-speed cruising. A 20% increase in top speed means a 40% increase in the energy to be dissipated in a crash stop and even this is a considerable underestimate of the extra load that can be thrown on the brakes by the use of full performance on ordinary roads. With 60% of the static weight on the front wheels, the front brakes must supply about 80% of the total stopping power for emergencies and from high speeds this is beyond the capabilities of the small drums that can be accommodated inside 10 in. rims, Lockheed has therefore developed special 7 in. discs for use on the front wheels which increase the rubbed friction area from 55 sq. in. to 120 sq. in. The pressure-limiting valve has been retained in the hydraulic circuit to the rear brakes, to stop them locking under very hard braking when forward weight transfer unloads the rear wheels, and a hydraulic pressure booster has been added in the line to the front discs. In effect this is the hydraulic equivalent of a two-speed gear which differentiates between the taking-up of pad clearances and light braking on the one hand and heavy braking on the other, automatically providing a greater leverage between pedal and brake pad for the heavy work without the excessive pedal travel that would otherwise be required for light conditions.

Externally the cars are almost indistinguishable from the standard models but new radiator grilles are used which differ for the Austin and Morris versions, both of which will be supplied in the same range of six duo-tone finishes. Inside the body a new instrument panel retains the large

A carpeted plywood floor covers the spare wheel and tools in the luggage locker of the Cooper and Super models. The number plate is hinged so that the car can be driven with the boot lid open.

AUSTIN 7 COOPER AND MORRIS MINI-COOPER

Engine	
Cylinders	Transverse 4 in line with 3-bearing crankshaft.
Bore and stroke ...	62.43 mm. × 81.28 mm. (2.458 in. × 3.20 in.).
Cubic capacity ...	997 c.c. (60.9 cu. in.).
Piston area	18.96 sq. in.
Compression ratio ...	9.0/1.
Valvegear	In-line vertical overhead valves operated by pushrods and rockers from chain-driven camshaft.
Carburation	Two 1¼ in. S.U. type HS2 inclined carburetters, fed by rear-mounted S.U. electric pump, from 5½-gallon tank.
Ignition	12-volt coil, centrifugal and vacuum timing control, 14 mm. Champion N5 sparking plugs.
Lubrication	Eccentric vane oil pump, Purolator full-flow filter and 8-pint sump (plus 1 pint in filter).
Cooling	Water cooling with pump, fan and thermostat; 5¼-pint water capacity.
Electrical system ...	12-volt 34-amp. hr. battery.
Maximum power ...	55 b.h.p. at 6,000 r.p.m., equivalent to 119 lb./sq. in. b.m.e.p. at 3,230 ft./min. piston speed and 2.9 b.h.p. per sq. in. of piston area.
Maximum torque ...	54.5 lb. ft. at 3,600 r.p.m., equivalent to 135 lb./sq. in. b.m.e.p. at 1,994 ft./min, piston speed.
Transmission	
Clutch	7⅛ in. single dry plate, hydraulically operated.
Gearbox	4-speed and reverse with direct top gear and synchromesh on upper 3 ratios.
Overall ratios ...	3.76, 5.11, 7.21 and 12.05; rev. 12.05.
Final drive	To front wheels via helical spur gears, universal joints and Birfield open drive shafts with Rzeppa constant velocity outer joints.
Chassis	
Brakes	Lockheed hydraulic with booster to front disc brakes, and drums at rear with pressure limiting valve.
Brake dimensions ...	Front discs 7 in. dia.; rear drums 7 in. dia. × 1¼ in wide.
Brake areas	45 sq. in. of lining (11 sq. in. front plus 34 sq. in. rear) working on 157.4 sq. in. rubbed area of discs and drums.
Front suspension ...	Independent by rubber cone spring units, unequal length transverse wishbones with outer ball joints and telescopic dampers.
Rear suspension ...	Independent by rubber cone spring units and single trailing arms; telescopic dampers.
Wheels and tyres ...	4-stud pressed steel wheels with 3.5 in. rims and 5.20-10 Dunlop nylon tyres.
Steering	Rack and pinion.
Dimensions	
Length	Overall 10 ft.; wheelbase 6 ft. 8 in.
Width	Overall 4 ft. 7 in.; track 3 ft. 11¾ in. at front and 3 ft. 10 in. at rear.
Height	4 ft. 5 in.; ground clearance 6 in.
Turning circle ...	31 ft.
Kerb weight	11¾ cwt. (without fuel but with oil, water, tools, spare wheel, etc.).
Effective Gearing	
Top gear ratio ...	14.9 m.p.h. at 1,000 r.p.m. and 27.4 m.p.h. at 1,000 ft./min. piston speed.
Maximum torque ...	3,600 r.p.m. corresponds to approx. 53.6 m.p.h. in top gear.
Maximum power ...	6,000 r.p.m. corresponds to approx. 89 m.p.h. in top gear.
Probable top gear pulling power... ...	245 lb./ton approx. (computed by *The Motor* from manufacturers' figures for torque, gear ratio and kerb weight, with allowances for 3½ cwt. load, 10% losses and 60 lb./ton drag).

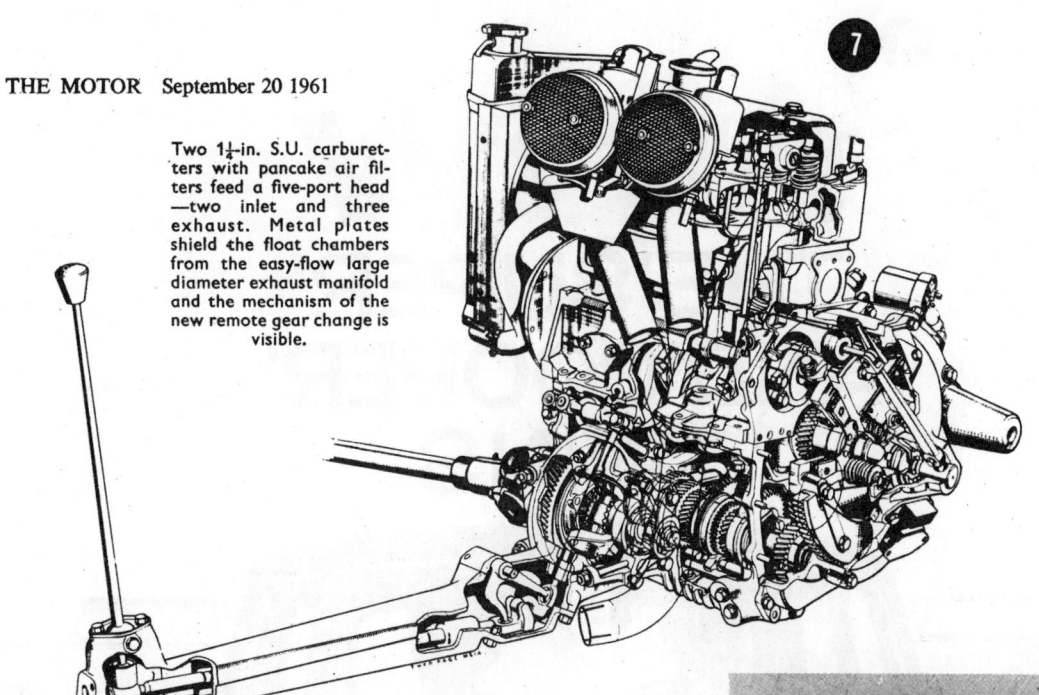

Two 1¼-in. S.U. carburetters with pancake air filters feed a five-port head —two inlet and three exhaust. Metal plates shield the float chambers from the easy-flow large diameter exhaust manifold and the mechanism of the new remote gear change is visible.

speedometer which still incorporates the fuel gauge but this is now flanked by separate oil pressure and water temperature gauges and an ignition key starter replaces the previous floor button. The front seats have been re-designed and the rear ones have higher backrests. A re-circulatory heater is part of the standard equipment on home models only but many people will be pleased to learn that there is full provision for the fitting of a fresh air type which will be available as an optional extra. Considerable attention has been paid to sound proofing, and damping materials have been used liberally on floor, roof and doors. Engine cooling fan noise is greatly reduced by a 16-bladed impeller driven at 1.2 times engine speed.

Like all other B.M.C. vehicles, the ADO 50 is fitted with built-in mounting points for seat belts which can be ordered as a standard accessory. It is interesting to note that the all-independent suspension of the standard car with its variable-rate rubber spring units has required no

modifications at all to make it suitable for a car with a performance equivalent to that of a more orthodox vehicle with an engine of twice the size.

. . . and a Super de luxe 850 Model

ANNOUNCED simultaneously with the B.M.C. Coopers are two special equipment versions of the standard car to be known as the Austin 7 Super and Morris Super Mini-Minor. These supplement the existing de luxe models, offering a higher standard of body trim, equipment and sound insulation. The mechanical specification is exactly the same as that of the standard 850 c.c. car but the bodywork and trim are almost identical with that of the B.M.C. Cooper and the cars are available in the same six duo-tone colours. The radiator grills have more vertical members than the standard ones.

The instrument nacelle has a combined

The cardioid combustion chamber, typical of current B.M.C. products directs incoming mixture towards the plug and away from the hot exhaust valve which would raise the charge temperature and reduce the volumetric efficiency. There is ample clearance between the chamber wall and the enlarged inlet valve so that flow is not restricted round half the periphery.

central speedometer and fuel gauge with a thermometer on one side, an oil pressure gauge on the other and a key-operated starter. The interior light is roof-mounted and the wire door pulls have been replaced by lever handles. A fitted carpet with heavy underfelt covers the whole of the interior floor and also the bottom of the boot whilst felt linings on the body panels and sound deadening emulsion on the wheel arches make further contribution to silence of travel.

Special seats have polyether foam cushions, squabs upholstered with rubberized horsehair and a covering of contrasting vinyl treated fabric. This washable material is also used for the rest of the interior trim including the windscreen frame and the edge of the parcel shelf. Standard equipment includes two sun visors and a recirculatory heater, a fresh-air heater being available at extra cost.

Prices of these new models had not been announced when these pages went to press.

An engine installation view shows the bulkhead sound insulating quilt, underbonnet insulation and on the left the fan and ducting for the optional fresh air heater.

NOW
THE 'SUPER' AND 'COOPER' TWINS

QUALITY FIRST

AUSTIN BMC THE BRITISH MOTOR CORPORATION LTD **MORRIS**

AUSTIN SUPER SEVEN ★ **MORRIS SUPER MINI-MINOR**	New styling, new luxury features everywhere. New design seats, softer furnishings. New interior door handles. Combined starter and ignition switch. Oil gauge and ammeter. New quiet ride. New stainless chrome door finish. Twice as many fashionable colours to choose from. Super-smart white wall tyres (extra).
AUSTIN SEVEN COOPER ★ **MORRIS MINI-COOPER**	Sports-car versions of the Austin and Morris 'Super' twins. Developed by B.M.C. in conjunction with John Cooper and warranted for 12 months by B.M.C. Sports-car specifications—the cars for *real* enthusiasts. Twin carburetters. Disc brakes. Speedo calibrated to 100 m.p.h. Remote control gear lever. Duo-tone colours

Twelve Months' Warranty and backed by B.M.C. Service
BRITAIN'S MOST CHALLENGING CARS
THE BRITISH MOTOR CORPORATION LIMITED BIRMINGHAM AND OXFORD

The Autocar road tests

1843

AUSTIN SEVEN COOPER

A die-cast intake grille, the Cooper name in the badge and the tubular end-trims on the bumper help to distinguish this from less potent Austin Sevens

CONCEIVED as a family four-seater with minimal exterior dimensions, the B.M.C. ADO15, by its good handling and rapid point-to-point capabilities in modern traffic conditions, has endeared itself to an additional section of the motoring public which would normally seek out a larger and more expensive car for its journeys. From this group of motorists, to whom expense within reason is not the main consideration, has come the main instigation for the performance improvement kits and conversions for Minis and other models which have appeared on the market in the last two years. It is natural, therefore, that the manufacturers themselves, seeing this potential market, should develop a model which would not only give improved performance but would combine with it accepted standards of reliability and, in view of the car's special character, a higher-than-normal level of finish as well.

The results are the Morris and Austin Coopers, so named because of the co-operation of Charles and John Cooper—of Cooper racing car fame—in their preparation. With an engine of 997 c.c. producing 55 b.h.p. at 6,000 r.p.m., in place of the standard 848 c.c. unit developing 37 b.h.p. at 5,500 r.p.m., disc brakes on the front wheels and many minor refinements, each is offered at a price of £679 7s 3d, approximately £126 more than the de luxe ADO15. The Austin version is the subject of this test.

Tractability was the immediate virtue noted with the car. Although it is said that lessons learned in formula Junior racing have been incorporated in the power unit, this has not been done at the expense of slow-speed running or middle-range performance. The best recorded speed on the road was 87 m.p.h., yet it was possible to pull away from as little as 14 m.p.h. in top gear without snatch. Indeed, once the novelty of the higher maximum speed has worn off, most owners will find that the main attraction of the car is its brisk performance in the important 40 to 70 m.p.h. range. The standing quarter-mile was covered in 20·9 seconds,

without recourse to top gear, and standstill to 60 m.p.h. took 18 seconds. With such a performance up its sleeve, undreamt of for a 1-litre saloon a few years back, the Austin Cooper becomes an astonishingly fast means of reaching B from A, whether it be Braemar from Axminster or Bayswater from Acton.

Two HS2 S.U. carburettors with pancake air filters supply mixture to the engine. Starting first thing in the morning is immediate with the use of the choke; after half a minute it can be pushed home, and thereafter during the day starts are possible without this aid. The water thermometer very quickly indicates normal running temperature and the engine runs sweetly during the warming-up period. There are none of the temperamental habits associated with tuned engines, apart from an uneven note on tick-over; neither in traffic nor in any other conditions encountered on the test was there a tendency to stall or overheat. After maintained high-speed running the engine did not lose its tune. The 9 to 1 compression ratio requires the use of super premium fuel which was consumed at the average rate of 27·14 m.p.g.

Retaining the compact overall dimensions of the type, the Austin Cooper is distinguished from the more mundane

Waterproofing of plugs, distributor and coil by plastic caps is standard practice. The optional fresh-air heater is seen to the left of the engine. General accessibility is still good, despite the well-packed appearance

Austin Seven
Cooper . . .

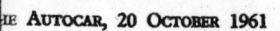

Tubular bumper-end additions are found also at the rear, where the Cooper name is given more prominence

models by a die-cast intake grille with horizontal slats; the name Cooper is incorporated in the radiator badge and in the boot lid motif. Curved, tubular chromium-plated additions at the bumper extremities, common to the Austin Super Seven, also identify the car. The interior trim of the test car was in red and grey plastic, with black leathercloth liberally applied to the windscreen surround. The padded edge of the parcels shelf was also covered in a similar black material, neatly finished with a chromium strip.

A central, oval instrument nacelle carries a speedometer without a trip recorder. This is combined with the fuel gauge and flanked by a water thermometer and an oil pressure gauge. The speedometer was unusually accurate, registering only 2 m.p.h. fast at 80 m.p.h.

Owing to the upright angle of the steering column, the two-spoke steering wheel is in a more horizontal plane than in most cars; in combination with the limited adjustment of the driving seat, this makes the driver feel somewhat restricted. However, a comfortable arm position is simply a matter of moving one's grip farther forward on the wheel. Once this is realized the driving position is, in fact, very comfortable, even on long journeys. Clutch and brake pedals are well spaced, and the throttle pedal is close enough to the front wheel-arch to allow this to be used as a foot steady. The vertical, remote-control gear-lever, although some distance from the steering wheel rim, is located ideally for quick changes; its action on the test car was slightly stiff, but precise, and the reverse guard-spring is strong enough to prevent inadvertent engagement of this gear.

All-round Visibility Good

All-round vision is excellent, although in a slightly lower plane than usual. There is adequate headroom above the front and back seats, and the top of the windscreen is just high enough to allow a full forward view without the driver having to duck his head. Both front wings are visible, and the rear window is virtually the extremity of the car, so judgment when reversing presents no problems. Because the roof line is low, entry into the car from a high pavement is difficult for elderly people, especially when trying to get to the back seat. However, once inside there is ample width across both front and rear seats and plenty of foot room.

The fresh-air heating system controls, the combined ignition-starter switch and toggle switches for the screen wipers and lamps are all arranged in a small horizontal panel in the middle of the parcels shelf. To the right of them is a screenwasher push-button, and to the left the choke control. They are neatly out of the way but the screenwasher could be closer to hand with advantage.

From the viewpoints of both traffic driving and performance, the choice of gearbox ratios could hardly be bettered.

The larger engine, giving 50 per cent more torque, has made higher indirect ratios practical; 46 m.p.h. is now the second gear maximum and almost 70 m.p.h. is possible in third. Since the engine will pull down to walking pace in second and almost that in third, most traffic driving is done in these ratios. The overall gearing of the Cooper models remains unchanged. Quick changes, made easier by the new remote lever, revealed shortcomings in the synchromesh when taking acceleration figures, but in normal use the slight delay essential to quiet engagement of the gears would not be irksome. Clutch action is both light and positive; with hard use there was no tendency to slip, but full depression of the pedal is essential to free the drive.

Engine tractability has been mentioned already; it is, indeed, extremely smooth throughout the range, and the 16-blade fan, now fitted to all ADO15s, is a tremendous improvement on the previous four-blader in respect of noise. Air intake noise is effectively silenced by the small air cleaners and considering the small amount of air space around the engine, the low level of noise transmitted to the interior of the car is a triumph for applied acoustics. For lazy driving one can travel mainly in top gear, but the power unit is so willing, and the torque from 1,500 r.p.m. upwards so even, that the temptation to accelerate hard in the indirects is hard to resist. The car will just pull away from a standstill on a 1-in-4 gradient.

Fuel consumption on this car depends more than usually on the way one drives it. As the constant speed figures show, an 80 m.p.h. blind down a motorway would be expensive, petrol being consumed at the rate of 23.3 m.p.g. However, a reduction of this cruising speed by only 10 m.p.h. reduces this figure to 32.8 m.p.g. The average owner should better 30 m.p.g. over a big mileage, even if reasonable use of the performance were made.

This particular car tested was one which had recently completed a high-speed run from Hamburg to Frankfurt, and though perfectly standard in other respects, was fitted

The rear number plate and lamp hinge down when the boot lid is opened. Beneath the wooden, carpeted floor is the battery and spare wheel

Austin Seven Cooper . . .

with twin fuel tanks. In normal trim, fuel capacity is 5·5 gallons, giving a touring range of, say, 150 miles, but fast motorway cruising would reduce the prudent distance between refills to about 100 miles. There is only a short filler neck to the tank, so the full flow of a service station pump can be taken without blow-back.

The superb handling of the ADO15s has become legendary in their two years of existence. As with all front-wheel-drive cars, the basic characteristic is that of understeer, but at slow cornering speeds this is negligible. Because of the Cooper's considerable extra power as compared with the standard Minis, the slip angle of its front tyres is greater through "full-throttle" corners; in other words, it understeers more. Consequently the sudden reduction in front tyre slip, if one closes the throttle in the middle of a corner, is much more abrupt and noticeable; in the same way one can "straighten out a corner" by opening the throttle wide in the middle of it. Given time and experience, the art of driving this car to its safe limits is soon acquired, because it is a vehicle which holds its driver's attention and interest through the sheer pleasure of driving it.

One of the main reasons for the extremely good road holding of this model is that there is very much more tyre

Two-tone plastic seat trim, door lock release lever and a B.M.C. standard seat belt. Left: Door panel trim and lining to the parcel pocket, together with the kick-plate, are among the interior refinements

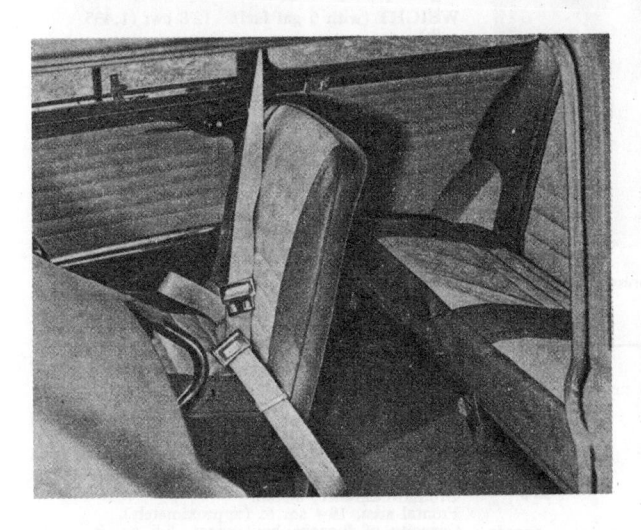

tread contact with the road than on most modern cars. Slight changes in tyre pressure will alter the handling of the car to a surprising degree. The good control of the Austin Cooper can be attributed partly to the precise and quick rack-and-pinion steering, which rarely betrays the fact that the front wheels are being driven.

No changes to either the rubber suspension or the shock absorber settings have been considered necessary for the Cooper models. There is that initial hardness in the ride which characterizes many Continental cars, but the progressive rubber suspension comes into its own as road surfaces deteriorate. The small wheels remain firmly in contact with the ground, the occupants following the car's abrupt vertical movements, as distinct from being bounced inside it. Driven two-up there is little roll during fast cornering, but with the rear seat occupied there is a tendency for the outside rear suspension to "sit down" on a sharp bend. A hand grip for the front passenger would add to his comfort and security in view of the limited lateral support given by the seats.

Over a very rough *pavé* test track, the suspension thudded noisily, but the car maintained an even keel and was controllable at speeds up to 40 m.p.h. On a washboard surface

the critical speed at which the suspension would iron out the irregularities was in excess of 50 m.p.h. Below this, vibration and discomfort increased as the speed dropped.

Quite one of the best features of the car is the Lockheed disc and drum brake system. Moderately high pedal pressures are required for small reductions in speed, but thereafter the action is progressive, with a reassuring feel. Pedal pressures of 100lb result in 90 per cent efficiency without wheel locking. Slightly higher pressures locked all four wheels evenly on a dry road. The brakes have a very reassuring capacity for stopping the car repeatedly from high speeds before any fade is induced, and recovery is rapid; they were practically unaffected by successive runs through a deep water splash. A strong arm is required to hold the car on a 1-in-4 gradient with the handbrake and there is a characteristic "unwinding" of the rear suspension as it is released.

Although the underbonnet space is well filled with carburettors and fresh-air heating unit, compared with the standard model, accessibility is not impaired. Waterproof covers are now fitted over the distributor cap and coil terminals. Service attention to 10 greasing points is required every 1,000 miles.

Luggage accommodation in the boot is adequate for a couple of medium-sized suitcases, leaving space to tuck away oddments and soft bags. However, this is not the whole story, for consideration must be given to the quite remarkable amount of stowage space inside the car. The large rear side pockets will take fairly large objects and there is room for two attaché cases beneath the rear seats; in addition the capacious containers in the doors will hold yet more impedimenta. On this model the boot has a carpeted wooden floor, with the spare wheel and battery beneath it. A meagre set of tools is supplied, including a screw jack which fits into a slot below the door and lifts the whole of one side of the car; it is best operated with the door open.

The fresh-air heating system, fitted as an extra, is a desirable accessory; it can eliminate completely the tendency for the windscreen to mist up in cold weather. Two pull knobs on the switch panel control, respectively, the amount of hot water going through the heater matrix, and the amount of fresh air admitted into the system. A flap on the heater outlet directs the whole output of the heater to the windscreen for extreme conditions; in the open position there is still an adequate air flow through the demister slots. A single-speed fan is provided to assist the flow in traffic conditions. The design of the fore-and-aft sliding side windows and the extractor-type at the rear allow an extremely good flow of fresh air through the car.

The flat driving mirror fitted does not take in the full view through the rear window. The non-parking screenwipers are efficient at speed but, due to the spindle location, leave an unswept area at the top right corner of the screen. On wet or dirty roads vision becomes obscured through the rear window, and when this happens the twin exterior rear-view mirrors are always a necessity.

Screenwipers and all other electrical equipment are to standard ADO 15 specification, save for the addition of a roof lamp with its own switch, and the deletion of the rear companion lamps. The double-dipping headlamps have separated, pre-focus bulbs, giving a good, long-range beam; when dipped they do not irritate oncoming drivers.

As road space in the British Isles, and elswhere, becomes more and more at a premium, small extra-performance cars like the Austin Seven Cooper will be in increasing demand. A significant sales factor will then be the assurance of the safety and durability of a product which has behind it the development resources of the original manufacturer, and is backed by his full guarantee. This latest B.M.C. product has these qualities in full.

AUSTIN SEVEN COOPER

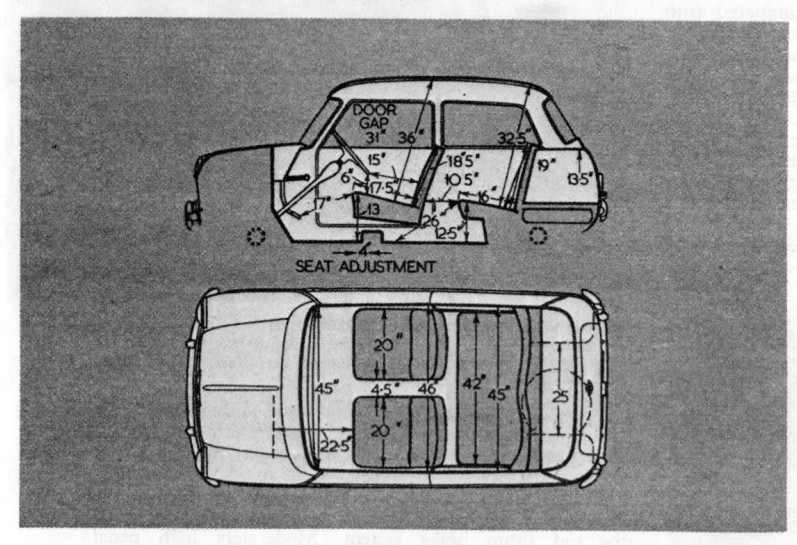

Scale ¼in. to 1ft. Driving seat in central position. Cushions uncompressed.

——— DATA ———

PRICE (basic), with saloon body, £465.
British purchase tax, £214 7s 3d.
Total (in Great Britain), £679 7s 3d.
Extras (inc. p.t.): Radio £30 3s 11d. Fresh-air heater £6 9s 6d. Safety belt £5 5s.

ENGINE: Capacity, 997 c.c. (60·8 cu. in.).
Number of cylinders, 4 in line.
Bore and stroke, 62·43×81·28mm (2·46 × 3·20in.).
Valve gear, overhead, pushrods and rockers.
Compression ratio, 9·0 to 1.
B.h.p. (net), 55 at 6,000 r.p.m. (B.h.p. per ton laden 69·5).
Torque, 54·5 lb. ft. at 3,600 r.p.m.
M.p.h. per 1,000 r.p.m. in top gear, 14·9.

WEIGHT (with 5 gal fuel): 12·8 cwt (1,435 lb).
Weight distribution (per cent): F, 60·9; R, 39·1.
Laden as tested, 15·8 cwt (1,771 lb).
Lb per c.c. (laden), 1·8.

BRAKES: Type, Lockheed hydraulic disc front, drum rear.
Drum dimensions: R, 7in. diameter; 1·25in. wide.
Disc diameter: F, 7in.
Swept area: F, 157 sq. in.; R, 55 sq. in. (268 sq. in. per ton laden).

TYRES: 5.20—10in. Dunlop Gold Seal with nylon casing.
Pressures (p.s.i.): F, 26; R, 24 (normal). F, 32; R, 30 (fast driving).

TANK CAPACITY: 5·5 Imperial gallons.
Oil sump, 8 pints.
Cooling system, 5·25 pints (plus 1 pint if heater fitted).

DIMENSIONS: Wheelbase, 6ft 8in.
Track: F, 3ft 11·75in.; R, 3ft 9·9in.
Length (overall), 10ft.
Width, 4ft 7in.
Height, 4ft 5in.
Ground clearance, 6·38in.
Frontal area, 15·4 sq. ft. (approximately).
Capacity of luggage boot space, 5·5 cu. ft.

ELECTRICAL SYSTEM: 12 - volt: 34 ampère-hour battery.
Headlamps, 50-40 watt bulbs.

SUSPENSION: Front, independent, wishbones and rubber cone springs.
Rear, independent, trailing arms with rubber cone springs.

——— PERFORMANCE ———

ACCELERATION TIMES:
Speed range, Gear Ratios and Time in Sec.

m.p.h.	3.77 to 1	4.68 to 1	7.22 to 1	12.04 to 1.
10—30	—	8·6	5·4	—
20—40	12·5	8·1	5·3	—
30—50	12·6	8·0	—	—
40—60	13·3	9·4	—	—
50—70	16·7	—	—	—

From rest through gears to:

30 m.p.h.	..	5.4 sec	
40 „	..	8·4 „	
50 „	..	12·6 „	
60 „	..	18·0 „	
70 „	..	27·5 „	
80 „	..	50·6 „	

Standing quarter mile 20·9 sec.

MAXIMUM SPEEDS ON GEARS:

Gear			m.p.h.	k.p.h.
Top	(mean)		84·7	136·2
	(best)		87·4	140·1
3rd	..	..	70·0	112·7
2nd	..	..	46·0	74·0
1st	..	..	28·0	45·1

TRACTIVE EFFORT (by Tapley meter):

			Pull (lb per ton)	Equivalent gradient
Top	..	..	220	1 in 10·1
Third	..	..	320	1 in 6·9
Second	..	..	470	1 in 4·7

BRAKES (at 30 m.p.h. in neutral):

Pedal load in lb	Retardation	Equiv. stopping distance in ft.
25	0·18g	168·0
50	0·49g	62·0
75	0·75g	40·5
100	0·90g	33·5

FUEL CONSUMPTION (at steady speeds in top gear):

30 m.p.h.	54.0 m.p.g.	
40 „	52·0 „	
50 „	45·4 „	
60 „	39·2 „	
70 „	32·8 „	
80 „	23·3 „	

Overall fuel consumption for 1,181 miles 26·8 m.p.g. (10·5 litres per 100 km.).
Approximate normal range 23-34 m.p.g. (12·3-8·3 litres per 100 km).
Fuel: Super Premium grade.

TEST CONDITIONS: Weather: Mainly dry, some showers. 5-8 m.p.h. wind.
Air temperature, 64 deg. F.
Model described in *The Autocar* of 22 September 1961.

STEERING: Turning circle:
Between kerbs: L, 33ft 1in.; R, 30ft 9in.
Between walls: L, 33ft 11in.; R, 31ft 8in.
Turns of steering wheel from lock to lock, 2·4.

SPEEDOMETER CORRECTION: m.p.h.

Car speedometer	..	10	20	30	40	50	60	70	80
True speed	..	11·5	20	29	39	48·5	58	68	78

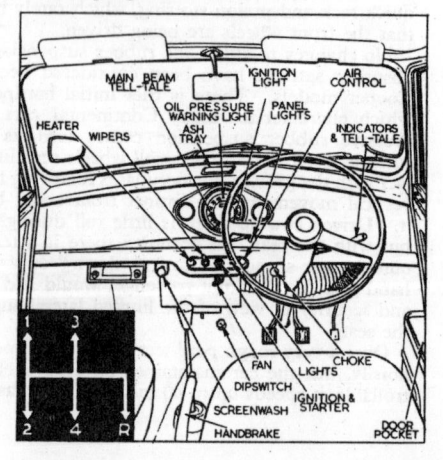

New grille and nameplate distinguish car, known as Austin 850 Cooper or Morris Mini-Cooper.

The more muscular Minis

A Lively Litre's Worth Of Fun That Rivals Many Sports Cars.   Report by David Phipps

The sports car is out of date. For some time the performance and controllability of small saloons have been almost comparable with those of open two-seaters. Now the saloon car, in the form of the Austin Seven Cooper and the Morris Mini-Cooper, is in the lead.

Superficially the ADO 50, as the new model is known at Cowley and Longbridge, is distinguishable from the standard ADO 15 only by two-tone paintwork (with a second colours for the roof) and a restyled grille. But take the car on the road and the difference is startlingly apparent. Even at tickover the engine betrays a more knobbly camshaft. Engage first gear, depress the little pedal on the right and this Mini fairly leaps away — with a screech of wheelspin — reaching 50 mph in just over 12 seconds (18 on the

Reworked engine features increased stroke, reduced bore, new cylinder head and two S. U. carburetors.

ADO 15) exceeding 70 mph in third gear and running happily up to a maximum of over 90 mph under favourable conditions. This alone could be considered impressive enough for an engine under 1 litre, but it is only on a twisty test course that the full potential of the car can be appreciated.

B.M.C. hold their new model demonstrations on a military vehicle test ground, using a 2-mile circuit which consists of two straights, two slightly banked curves and a series of undulating corners appropriately named the Snake Section. When the ADO 15 was first shown, two years ago, it impressed greatly by its road-holding and handling, but seemed underpowered and wrongly geared on the Snake Section. By comparison, the ADO 50 was a revelation. It howled through the tight Ess-bend at the end of the straight and held an indicated 50 mph in second gear through the long hairpin which followed. Over a blind crest, into third, and through the next left-hander the little car could even be persuaded to lift the inside front wheel — though with no apparent adverse effects.

And so through another right and left and on to the straight, after which the Mini happily went "flat" through the first banked turn — building up speed all the time — and through the gentler curve at the end of the main straight. Firm braking at 100 yards dissipated the speed in ample time for the Snake Section.

Such performance, from what is basically one of the world's cheapest cars, is quite staggering, especially as it is provided by the manufacturer and not by a specialist tuning concern. How has it been done? The 17 per cent increase in capacity has been achieved, somewhat surprisingly, by increasing the stroke (68.26 mm to 81.28 mm) and reducing the bore (62.9 mm to 62.43 mm). Improved gas flow is provided by a new cylinder head, with larger inlet valves and ports, modified combustion chambers and double valve springs. A new inlet manifold carries two S.U. carburetors and a larger diameter exhaust system is fitted. Compression ratio is raised from 8.3 to 9 to 1 and the new camshaft, mentioned above, increases valve overlap from 15 degrees to 37 degrees (the timing is, in fact, the same as on the MGA 1600 Mark Two). Power output is raised from 34 bhp at 5500 rpm to 55 bhp at 6000 rpm.

To match the increased performance, 7 inch Lockheed disc brakes are fitted at the front, in conjunction with 7 inch rear drums, operated through a pressure-limiting valve which prevents inadvertent locking. No chassis or suspension modifications have been made, but — to judge from the car's behaviour on the test course — none are necessary.

Inside the car the most noticeable change is the fitting of a remote control gear-shift. In theory this should provide a much better change than the long shift lever of the standard model, but on the two cars I tried the change from second to third was very imprecise, the lever preferring to go into a neutral position beyond third and at the back of reverse. This apart, all the controls are a delight to use. The steering is beautifully light and direct, and understeer at high speeds can always be encountered by lifting off slightly. The clutch and brake pedals require only light pressure, and it is perfectly easy to heel-and-toe, despite the tiny throttle pedal. The original single instrument (speedometer and fuel gauge) is now augmented by oil pressure and water temperature gauges. Interior trim is improved, with a new type of fabric for the seats, and a Fresh-Air heater is available.

The performance of the ADO 50 can be summed up by comparing it with a 2½-litre compact. In a straight line the acceleration and maximum speed of the two cars are almost identical. But on a twisty road, or something in the nature of an Alpine pass, there is no longer any comparison. And apart from being more roadworthy, and so much more fun to drive, the ADO 50 is also more comfortable — at least for this 6 ft 5 in. driver. It actually provides almost as much interior space as the compact, and at something like half the price. A few days ago a Frenchman asked me, as I stepped from my Standard Austin 850, "How does such big man get in such small car?" My answer, "Because it is really a big car," obviously baffled him. But with 55 bhp and disc brakes the ADO 50 is now a big car in every sense of the word, and I just can't wait to get hold of one.

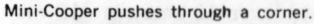
Mini-Cooper pushes through a corner.

IMPROVING THE PERFORMANCE OF POPULAR CARS

ALEXANDER MINI-COOPER

THERE is no denying the remarkable potentialities of the ADO 15. Although basically a small, inexpensive car for the masses, with suitable "tweaking" it is capable of outpacing many sports cars, particularly if the 997 c.c. Cooper version is used as a basis.

Recently a modified Morris Mini-Cooper was offered for test by Alexander Engineering. This had their "Big Bore" conversion in which the cylinders had been opened out to give a total displacement of 1,122 c.c. Special forged pistons, developed in Germany, and racing quality main and big-end bearing shells were fitted. Valve gear modifications consist of a new camshaft with extended valve opening periods, inlet and exhaust valves with heads of larger diameter and double springs.

Combustion chambers and ports had been extensively reshaped and polished, the compression ratio being raised from 9.0 to 9.5 to 1. Two 1.5in. dia. S.U. carburettors were mounted on a modified inlet manifold and the three-branch exhaust manifold was fabricated from tube.

Jet levers on the carburettors had not been connected to a facia control but the engine always started readily, though it would not idle until warm. Some extra work had been carried out on this engine to balance, within closer limits, the crankshaft assembly and pistons and there was no vibration period through the speed range up to a little over 7,000 r.p.m., at which maximum power is developed. The recommended engine speed limit is 7,500 r.p.m., corresponding to speeds in the indirect gears of 35, 59 and 83 m.p.h. but there was no valve bounce when nearly 8,000 r.p.m. was recorded by the Smiths electric tachometer which was fitted.

From rest through the gears outstanding improvements in acceleration times were obtained, compared with the standard Mini-Cooper. For example, the time to 60 m.p.h. was reduced from 18sec to 10.7sec and to 80 m.p.h. from 50.6sec to 19.9sec. Over 100 m.p.h. was reached without difficulty, the best figure recorded being a true 102.

This car did not have a particularly noisy exhaust, and 80-85 m.p.h. could be held for cruising without the engine becoming too obtrusive. However, above 4,500 r.p.m. the gear lever vibrated badly; this could be checked by placing a hand on it.

Just visible beneath the oversize 1.5in. dia. S.U. carburettors is the three-branch fabricated exhaust manifold

In town this Mini is docile and tractable and it is possible to pull away smoothly from 15 m.p.h. in top (1,000 r.p.m.). Between 2,000 and 3,000 r.p.m. irregular carburation on large throttle openings spoilt those acceleration figures in the gears which included this engine speed range, but with a light throttle this trouble did not appear. Carburettor needle profiles are to be altered to overcome this. There is plenty of torque at 3,000 r.p.m. and at about 3,800 r.p.m. this engine really takes hold. We were advised to use super premium fuel and did so for all performance testing, but it also ran well on premium, pinking only if allowed to labour.

As expected, road holding and handling were well able to cope with the extra performance. Koni dampers, with stiff settings for competition work were fitted. The brakes, however, with standard pads and linings, were barely up to the task, and at lower speeds a high pedal effort was necessary.

Performance Data

Figures in brackets are for the Austin Mini-Cooper tested in Autocar of 20 October 1961

Acceleration times (mean): *Speed range, Gear ratios, and Time in seconds:*

m.p.h.	3.77 to 1	5.11 to 1	7.21 to 1	12.05 to 1
10-30	—	8.5 (8.6)	5.2 (5.4)	3.5 (—)
20-40	13.7 (12.5)	8.0 (8.1)	4.5 (5.3)	—
30-50	15.2 (12.6)	6.9 (8.0)	4.7 (—)	—
40-60	11.9 (13.3)	7.2 (9.4)	5.6 (—)	—
50-70	11.0 (16.7)	7.8 (—)	—	—
60-80	13.9 (—)	9.2 (—)	—	—
70-90	20.8 (—)	—	—	—

From rest through gears to:

30 m.p.h.	..	..	3.5 sec.	(5.4)	sec.
40 ,,	..	..	5.4 ,,	(8.4)	,,
50 ,,	..	..	7.6 ,,	(12.6)	,,
60 ,,	..	..	10.7 ,,	(18.0)	,,
70 ,,	..	..	15.1 ,,	(27.5)	,,
80 ,,	..	..	19.9 ,,	(50.6)	,,
90 ,,	..	..	29.7 ,,	(—)	,,

Standing start quarter-mile 18.2 sec. (20.9)

Maximum Speed:

				m.p.h.	k.p.h.
Top (mean)	..	..	..	100.5 (84.7)	161.7 (136.2)
(best)	..	..	..	102.0 (87.4)	164.1 (140.1)
3rd	..	..	..	83.0 (70.0)	133.0 (113.0)
2nd	..	..	..	59.0 (46.0)	95.0 (74.0)
1st	..	..	..	35.0 (28.0)	56.0 (45.0)

Overall fuel consumption for 270 miles: 24.8 **m.p.g.,** 11.4 litres per 100 km. (37.0 m.p.g., 7.6 litres per 100 km.).

Prices:
Kit of parts on exchange basis and boring out cylinder block £125. Installed at works:
Labour, stripping and rebuilding engine, £12 10s.
Removing and refitting engine, £5 10s.
Balancing crankshaft assembly (optional), £15.

Alexander Engineering Co. Ltd., Haddenham, Buckinghamshire. Tel.: Haddenham 345-6.

JOHN BOLSTER

TESTS THE

MORRIS MINI-COOPER

THE Morris Mini-Minor has had a phenomenal success. One merely has to count them on the road to realize that Alec Issigonis has really hit the jackpot with this model. At a basic price equivalent to about a pre-war £100, it provides rapid, economical transport for four people with a standard of roadholding that confers quite exceptional safety on all road surfaces.

The basic design, with all four wheels independently sprung on rubber, is obviously suitable for considerably higher speeds than the standard 850 c.c. engine can provide. Indeed, many tuned versions of

this power plant prove the point. It was therefore decided to produce a faster version of the Mini, and Alec Issigonis went into a huddle with John Cooper over this.

In order to keep the price within bounds, it was decided to use the same pressed-steel body. The engine is still transversely mounted and drives the front wheels, but it has a crankshaft giving a longer stroke which has a vibration damper for sustained operation at high revolutions. The capacity of this engine is in fact slightly greater than that of the old Minor 1000 at 997 c.c. and there are twin SU carburetters. This larger

power unit, with its special camshaft, adds no less than 21.5 b.h.p. to the Mini.

The unit is by no means highly tuned, but it does its best work at about 1,000 r.p.m. above the usual rate of the smaller version. It therefore needs closer gear ratios which are provided, the lower gears having been "closed up", but the final drive (top gear) ratio is the same, though an optional "high cog" is available. The brakes of the ordinary Mini are not its strongest feature, so discs are installed on the front wheels of the Cooper.

Apart from more complete instrumentation, as befits a sporting type of car, the interior does not differ greatly from that of the standard model. The most noticeable difference is a remote control gear lever which is a reasonably effective makeshift.

The Mini-Cooper can be used for any purpose for which the standard model is suitable. It is just as good a shopping car and not noticeably noisier than its bread-and-butter sister. In this connection, however, one must criticize the remote control gear lever, for it does "telephone" a lot of noise into the interior of the vehicle. The actual changes are quite quick, and the closer ratios greatly improve the car, second now becoming a really useful gear.

The clutch grips well for fast changes, and one must applaud the responsiveness of all the controls. The ride is fairly hard, with some pitching, but many sports car enthusiasts are by no means averse to such suspension characteristics. The standard of comfort is acceptable, and in the case of the test car it was greatly enhanced by the optional fresh air heater, a tremendous improvement.

The speed of the Mini-Cooper is held down to a little below 90 m.p.h. by its rather unstreamlined shape. Tuned Minis have certainly gone faster, but the extra power needed to increase the speed appreciably could only be supplied by a rather "hot" and not very economical engine. The acceleration is in a different world from that of the standard Mini, and cars which habitually overtake that worthy little machine are themselves overwhelmed by the Cooper version.

The average speeds which can be achieved, particularly over difficult terrain, can only be described as incredible. The engine is very willing and the gear ratios are so right that even very fast sports cars cannot shake off this Mini. The disc brakes, which initially left something to be desired, are now perfectly adequate and really pin the little projectile down.

Nevertheless, it is the celebrated road-holding which contributes most to the overall performance. We have all seen Minis in saloon car races demonstrating their high cornering power against more conventional cars. Their cornering speed is certainly a little greater than that of most of their competitors. What is so remarkable, though, is the phenomenal "dicing margin" that is available.

Most cars with high cornering power tend to be unforgiving. In the hands of an expert, they are most impressive, but the novice who tries to drive on the limit will eventually spin off ignominiously. The Mini-Cooper can be driven up to and past the limit of adhesion by quite a moderate driver. When he appears to be about to enter the *decor* he simply eases his foot momentarily. The tail comes round, the sliding car loses speed, and another burst of throttle sends him on his way.

Fundamentally, this type of stability renders it possible to travel fast in safety even when the road is not well known to the driver. He will find that he can beat pilots of his own calibre who have faster cars, simply because his Mini-Cooper will look after him when he is indiscreet. The extra power available makes the Cooper version even safer than the standard car.

The only danger is that the Mini driver may later try to handle something else with similar carefree abandon. Frankly, it can't be done!

It is natural that the extra performance must be paid for in some way. The fuel consumption is considerably heavier than that of the standard Mini, which habitually achieves well over 40 m.p.g. If one uses the performance, the Cooper will stay on the wrong side of 30 m.p.g., but this is entirely reasonable at the average speed which it encompasses. Similarly, the driver who habitually corners near the limit will consume his tyres fairly rapidly and they become remarkably hot during long runs at spectacular average speeds.

The Mini-Cooper is a small economy saloon when driven moderately, which is flexible and easy to handle. When pressed, it becomes a genuine sports car, capable of really remarkable performances. If it is then somewhat less economical, the sheer fun of handling it must more than counter-balance the slight increase in cost. For the man who is bored with his daily drive to the office, a Mini-Cooper could render this dreary trip an eagerly awaited pleasure.

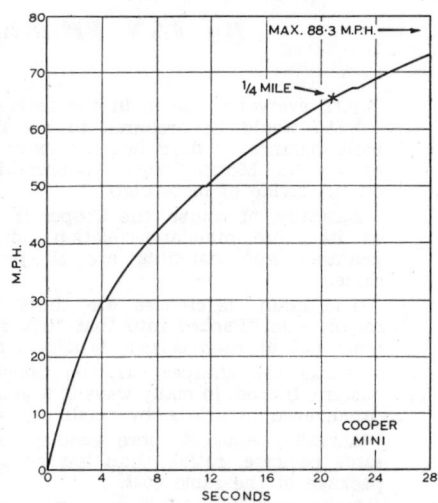

ACCELERATION GRAPH

SPECIFICATION AND PERFORMANCE DATA

Car Tested: Morris Mini-Cooper saloon, price £640 including P.T.

Engine: Four-cylinders 62.43 mm. × 81.33 mm. (997 c.c.). Push-rod-operated overhead valves. Compression ratio 9 to 1. 56 b.h.p. (DIN) at 6,000 r.p.m. Twin SU carburetters. Lucas coil and distributor.

Transmission: Single dry-plate clutch. Four-speed gearbox with central remote control and synchromesh on upper three gears, ratios 3.765, 5.109, 7.214, and 12.048 to 1. All-gear transmission to differential and articulated shafts to front hubs.

Chassis: Steel punt-type chassis with integral steel saloon body. Independent front suspension by wishbones and rubber springs. Rack and pinion steering. Independent rear suspension by trailing arms and rubber springs. Telescopic dampers all round. Bolt-on disc wheels fitted 5.20-10 ins. tyres. Lockheed hydraulic brakes with front discs and rear drums.

Equipment: 12-volt lighting and starting. Speedometer. Temperature, oil pressure, and fuel gauges. Windscreen wipers and washers. Heating and demisting. Flashing direction indicators.

Dimensions: Wheelbase 6 ft. 8⅜ ins. Track (front) 3 ft. 11⅚ ins.; (rear) 3 ft. 9⅞ ins. Overall length 10 ft. 0⅛ in. Width 4 ft. 7½ ins. Turning circle 31 ft. Weight 11½ cwt.

Performance: Maximum speed 88.3 m.p.h. Speeds in gears: 3rd, 67 m.p.h.; 2nd, 50 m.p.h.; 1st, 30 m.p.h. Standing quarter-mile 20.8 s. Acceleration: 0-30 m.p.h., 4 s.; 0-50 m.p.h., 11.2 s.; 0-60 m.p.h., 17 s.; 0-70 m.p.h., 25.1 s.

Fuel Consumption: 27 to 31 m.p.g.

Where the £200 goes

IS THE MINI-COOPER TOO COSTLY?

Indecently fast for one so small, BMC's Cooper-Mini is such a projectile that few vehicles, including sports cars, match its performance.

By IAN FRASER

Radiator grille of the Cooper-Mini — instantly distinguishing feature. Behind that grille, in the engine compartment, is the car's success story.

NOT everyone who is in the market for a Morris 850 would be prepared to dig deeper into his bank balance to drag out the extra £200 necessary to buy the Morris-Cooper (Cooper-Mini, if you are on the inside of motoring.)

Basically, of course, the Cooper is much the same as its more mundane relation; size, general appearance and handling are all pretty much the same.

The main differences are those which are not readily seen. Packed into that 10 ft ½ in box, with a doughnut at each corner, is a real motor car.

Unlike the cheaper car, the Cooper is far from austere. Indeed, in many ways is it genuinely sophisticated, even luxurious, by small car standards.

Actually, you get more goodies packed into the same package, rather than less goodies in a bigger package of the same cost.

It is easy to pick out a Cooper on the road, however. The radiator grille is much bolder and the bumpers have small, tubular steel extensions, presumably to protect the vehicle from the ravishes of parking among normal size and big cars. To me they looked far from able to repel the onslaught of a "heavy".

Clue number three to identity is the slotted chrome wheel disc set and the two tone paintwork which seems to have more lustre than the 850's. Failing all this, you can read the nameplates.

While the exterior is very Morris 850, the interior is not. When you open a door the interior almost comes out to meet you. The seats are trimmed in a beautifully smooth and soft plastic material that looks and feels like leather, rather than that mottled (ugh) sloppy covering of the 850. Both front and rear seat shapes are much the same as the cheaper car's, but the trim itself transforms the appearance.

On the floor there is carpet back and front. It is not really good carpet, but at least it is better than rubber matting moulded to look like something it is not and never will be.

The neat, remote control gear shift lever sticks up from the floor just forward of the handbrake, which is between the seats. Pedals — pendant type — look small in comparison to the width of the front compartment.

More instruments than the sparse collection in the 850 grace the dash. The speedo, neatly calibrated to 100 mph also includes the fuel gauge, but the nacelle has been extended to include oil pressure and water temperature gauges. Instead of the painted metal, the scuttle and bottom part of the commodious parcel tray are covered in black vinyl material to cut screen reflections.

Rather jarring are the controls for the heating system — standard equipment, by the way — which are pulled out to switch off the warm air. Ungainly to my eyes, it may have advantages in England and Europe where heaters are used most of the time.

A radio was in the test car, but since no provision has been made for fitting one, it has to reside under the tray on the passenger's side, rendering it impossible to operate from the driver's seat if a safety belt is being worn.

There is an interior light. Its operation is purely manual, which seems a pity. Also the screen wipers must be parked manually — strange omissions for a car that is trying hard to forget its more mundane heritage.

WHEELS FULL ROAD TEST

Special wheel discs, two-tone paintwork, tubular bumper extensions and a big diameter exhaust pipe.

Attached to the front end of this package is the combined engine/transmission unit. Actually, it bears only a little resemblance to that of the 850. The cylinders have longer strokes and smaller bores to achieve a cubic capacity of 997 cc, the materials used in the crankshaft and bearings are more suitable for high performance engines. Although basically the BMC A-series cylinder head, it does differ to make it breathe more freely. Both induction and exhaust systems are totally different, the compression ratio is 9 to 1 and camshaft is decidely sporty.

The close ratio gearbox (the parts are available for 850's) lives down in the sump. Unfortunately, BMC designer's did not think it necessary to improve the synchromesh on the Cooper, although the models announced in England at motor show time have the Morris 1100's baulk-ring synchromesh.

The test car had done nearly 2000 miles when we picked it up, but evidence of the type of miles was the fact that synchromesh on second was almost completely inoperative. We experienced difficulty in engaging either first or second from a standing start on a number of occasions — a problem apparent in a number of BMC products.

Against this, the ratios of the Cooper are very well chosen and match the power curve nicely. When it is necessary to engage first on the move, double de-clutching allows easy and silent selection — easier, perhaps, than with any other BMC product.

Of course, it is not necessary to whack the lever back to low frequently since the engine is sufficiently flexible for reasonably smooth starts to be made in second gear. For really snappy results out of uphill hairpin bends, first is essential.

Interior carpets have materially assisted in reducing the mechanical noises that are inclined to make one's ears ring in the 850. Exhaust noise is surprisingly low considering the size of outlet pipe. Even outside the car it is modest.

The interior has not been overlooked in the Cooper. There are two additional instruments. For heater operation, knobs must be pushed in.

The real secret of the Cooper-Mini's success is not in its refinement or prestige value, but in plain hairy performance.

When you first leap enthusiastically in behind that oversized steering wheel, get the motor started and move off there is an anti-climax. Things don't feel much different to a standard 850.

You realise this car is a "full bottle" only when the accelerator is flattened to the floor. In second, for instance, this does not bring a startling response up to 25 mph, but from there to an indicated 50 mph the effect is shattering. The important thing is to treat the markings on the speedo seriously and let the engine wind out (to about 6000 rpm, I should think), regardless of the gearlever's rattling.

A staggering 55 bhp has been extracted from the 1-litre engine. It runs on a 9 to 1 compression, spins to considerably more than 6000 rpm.

Pretty much the same kind of thing happens in any gear, although second is my choice for enterprising travel around the city and suburbs. A satisfactory technique in my book is to get the Cooper well and truly rolling in first, change into second and wind it up to a useful suburban cruising speed before selecting top, giving third a miss.

Out on the open road, all the gears are there to use. Third is a very good overtaking ratio with a maximum speed in the 70's. Not that top is sluggish by any means, but it does come into its own from 60 upwards.

Naturally, you don't get all this performance for nothing. The Cooper wolfs down more petrol than the 850, but even a lead-foot would be hard pressed to get less than 32 mpg.

Handling is very Morris 850. It understeers quite strongly with the power on, but this turns to strong oversteer in a snap if the power is cut off.

This is the one trap with the Cooper. It has enough power available at the front wheels to get the car around corners very quickly indeed. If conditions should become unfavorable and power had to be cut off, I venture to say it would take a remarkably nimble driver to catch the tail before the whole situation gets completely out of hand. For this reason the Cooper can be difficult, particularly in the hands of a person who has had little or no experience with front wheel drive. Perhaps it would be at its best with a graduate Morris 850 driver behind the wheel — one who has had the benefit of a few good scares.

For a car with this one's potential, I was not wildly enthusiastic about the brakes. There are tiny discs on the front and drums on the rear with pressure limiting devices to prevent wheel locking.

Not unexpectedly, some fade was experienced during our tests, but was overcome by leaning even harder on the pedal. At low speeds around town it is necessary to really stand on the postage-stamp sized brake pedal to get an effectively rapid stop. The handbrake was so effective and the Dunlop nylon tyres so adhesive that it was possible to drive away, one person up, with the back wheels locked solid.

There is no doubt that the Morris-Cooper is a highly desirable piece of work for around town and not-too-distant country trips (the driving position is wearing for a tall person).

Value for money there is nothing with four seats that even looks at a Cooper's £950 price tag. Significantly, it is noticably faster than BMC's comparable sports car, the Austin Healey Sprite.   #

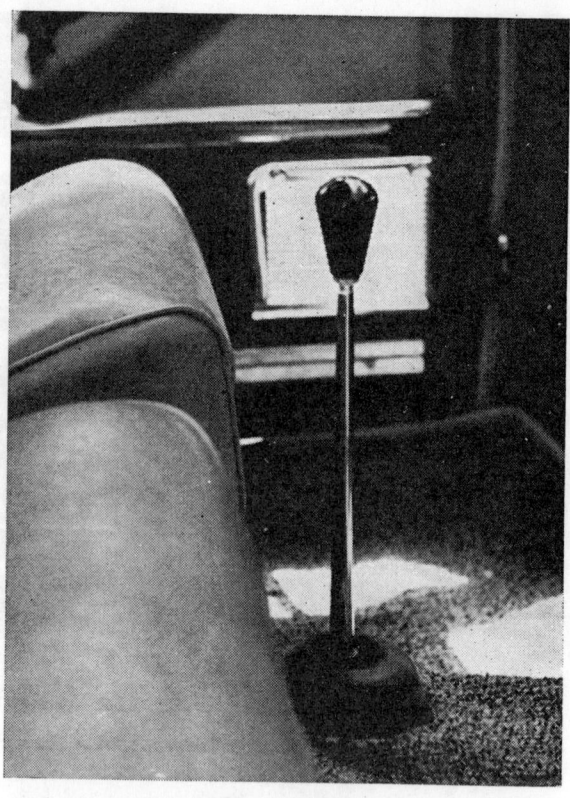

Remote control gearshift on the floor works much better than the direct lever on the 850. Note carpeted floors.

Tubular extension on the bumper bars may help prevent body damage by big cars, but would not be of great assistance.

wheels ROAD TEST

TECHNICAL DETAILS

OF THE

MORRIS-COOPER

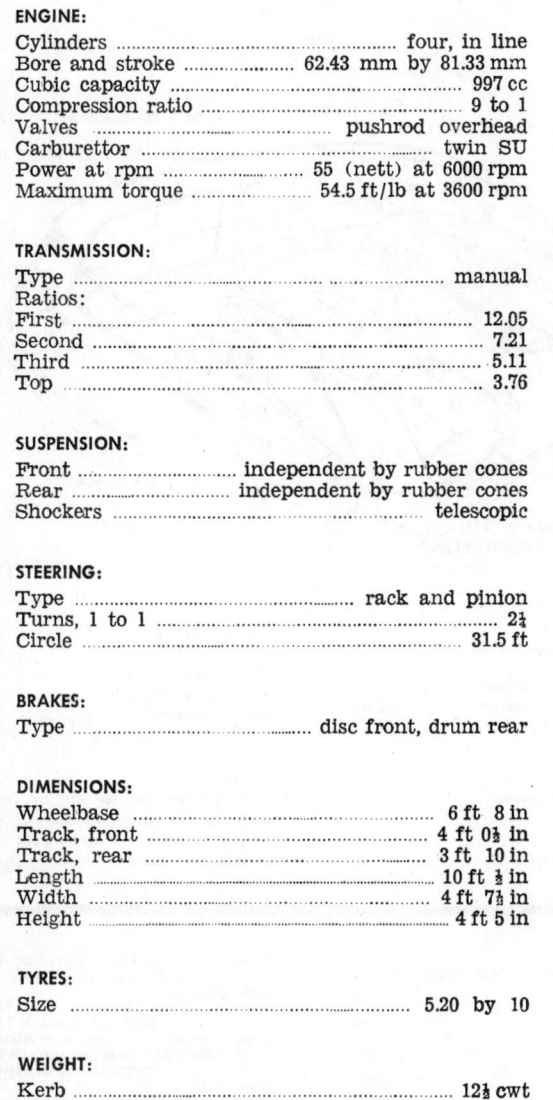

SPECIFICATIONS

ENGINE:
Cylinders .... four, in line
Bore and stroke .... 62.43 mm by 81.33 mm
Cubic capacity .... 997 cc
Compression ratio .... 9 to 1
Valves .... pushrod overhead
Carburettor .... twin SU
Power at rpm .... 55 (nett) at 6000 rpm
Maximum torque .... 54.5 ft/lb at 3600 rpm

TRANSMISSION:
Type .... manual
Ratios:
First 12.05
Second 7.21
Third 5.11
Top 3.76

SUSPENSION:
Front .... independent by rubber cones
Rear .... independent by rubber cones
Shockers .... telescopic

STEERING:
Type .... rack and pinion
Turns, 1 to 1 .... 2¼
Circle .... 31.5 ft

BRAKES:
Type .... disc front, drum rear

DIMENSIONS:
Wheelbase .... 6 ft 8 in
Track, front .... 4 ft 0½ in
Track, rear .... 3 ft 10 in
Length .... 10 ft ⅛ in
Width .... 4 ft 7½ in
Height .... 4 ft 5 in

TYRES:
Size .... 5.20 by 10

WEIGHT:
Kerb .... 12½ cwt

PERFORMANCE

TOP SPEED:
Fastest run .... 86.1 mph
Average of all runs .... 83.2 mph

MAXIMUM SPEED IN GEARS:
First .... 30 mph
Second .... 48 mph
Third .... 67 mph
Top .... 86.1 mph

ACCELERATION:
Standing Quarter Mile:
Fastest run .... 20.9 sec
Average of all runs .... 21.2 sec
0 to 30 mph .... 4.7 sec
0 to 40 mph .... 7.5 sec
0 to 50 mph .... 11.8 sec
0 to 60 mph .... 16.8 sec
0 to 70 mph .... 26.9 sec
0 to 80 mph .... NA sec
20 to 40 mph .... 11.7 sec
30 to 50 mph .... 12.3 sec
40 to 60 mph .... 12.9 sec

GO-TO-WHOA:
0-60-0 mph .... NA

SPEEDO ERROR:

Indicated	Actual
30 mph	29 mph
40 mph	39 mph
50 mph	48 mph
60 mph	58 mph
70 mph	67 mph
80 mph	NA
90 mph	NA

FUEL CONSUMPTION:
Cruising speed .... 37 mpg
Overall for test .... 27 mpg

THE Motor

MAKE: Morris *TYPE: Mini-Cooper S*

MAKERS: British Motor Corporation Limited, Longbridge, Birmingham

ROAD TEST ● No. 16/63

TEST DATA:

CONDITIONS: *Weather: Mild with light wind (10 m.p.h.) and intermittent rain. (Temperature 45°-50°F., Barometer 29·8 in Hg.) Surface: Damp during acceleration tests; otherwise dry. Fuel: Premium grade pump petrol (98 Octane by Research Method).*

MAXIMUM SPEEDS

Mean lap speed around banked circuit	94·5 m.p.h.
Best one-way ¼-mile time equals	98·9 m.p.h.

"Maximile" Speed (Timed quarter mile after one mile accelerating from rest)

Mean of opposite runs	91·8 m.p.h.
Best one-way time equals ..	93·8 m.p.h.

Speed in gears

Max. speed in 3rd gear	84 m.p.h.
Max. speed in 2nd gear	62 m.p.h.
Max. speed in 1st gear	37 m.p.h.

ACCELERATION TIMES from standstill

0-30 m.p.h.	4·0 sec.
0-40 m.p.h.	6·9 sec.
0-50 m.p.h.	9·0 sec.
0-60 m.p.h.	12·9 sec.
0-70 m.p.h.	17·1 sec.
0-80 m.p.h.	23·2 sec.
0-90 m.p.h.	40·1 sec.
Standing quarter mile	18·9 sec.

ACCELERATION TIMES on upper ratios

	Top gear	Third gear
10-30 m.p.h.	11·6 sec.	7·9 sec.
20-40 m.p.h.	10·4 sec.	6·7 sec.
30-50 m.p.h.	11·0 sec.	7·0 sec.
40-60 m.p.h.	10·8 sec.	7·2 sec.
50-70 m.p.h.	12·1 sec.	8·2 sec.
60-80 m.p.h.	16·0 sec.	10·7 sec.
70-90 m.p.h.	26·4 sec.	—

HILL CLIMBING

Max. gradient climbable at steady speed.

Top gear	1 in 10·1 (Tapley 220 lb./ton)
3rd gear	1 in 6·4 (Tapley 345 lb./ton)
2nd gear	1 in 4·4 (Tapley 495 lb./ton)

FUEL CONSUMPTION

Overall Fuel Consumption for 1,345 miles, 50½ gallons, equals 26·8 m.p.g. (10·55 litres/100 km.)

Touring Fuel Consumption (m.p.g. at steady speed midway between 30 m.p.h. and maximum, less 5% allowance for acceleration) 39·5 m.p.g.

Fuel tank capacity (maker's figure) 5½ gallons

Direct top gear

57½ m.p.g.	at constant 30 m.p.h. on level
54½ m.p.g.	at constant 40 m.p.h. on level
50½ m.p.g.	at constant 50 m.p.h. on level
42½ m.p.g.	at constant 60 m.p.h. on level
38½ m.p.g.	at constant 70 m.p.h. on level
32 m.p.g.	at constant 80 m.p.h. on level
21½ m.p.g.	at constant 90 m.p.h. on level

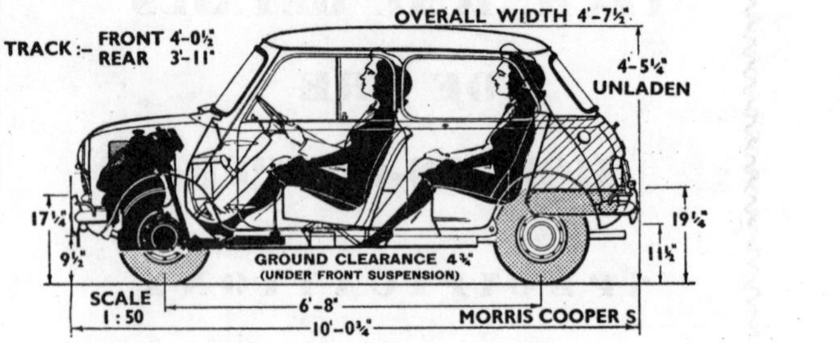
OVERALL WIDTH 4'-7½"
TRACK:- FRONT 4'-0½" REAR 3'-11"
4'-5¼" UNLADEN
17¼"
9½"
GROUND CLEARANCE 4¾" (UNDER FRONT SUSPENSION)
19¼"
11½"
SCALE 1:50
6'-8"
10'-0¾"
MORRIS COOPER S

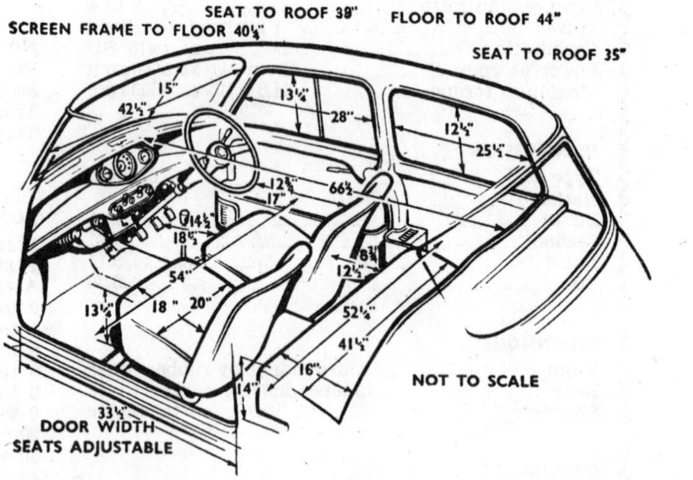
SEAT TO ROOF 38"
FLOOR TO ROOF 44"
SCREEN FRAME TO FLOOR 40¼"
SEAT TO ROOF 35"
15"
42½"
13¼"
28"
12½"
25½"
12¾"
17"
66½
14½"
18½"
8¾"
12½"
54"
13¼"
18"
20"
52¼"
41¼"
16"
14"
33½"
DOOR WIDTH
SEATS ADJUSTABLE
NOT TO SCALE

BRAKES

Deceleration and equivalent s opping distance from 30 m.p.h.

1·00 g with 60 lb. pedal pressure ..	(30 ft.)
·71 g with 50 lb. pedal pressure ..	(42 ft.)
·37 g with 25 lb. pedal pressure ..	(81 ft.)

STEERING

Turning circle between kerbs:

Left	31¾ft.
Right	29 ft.
Turns of steering wheel from lock to lock	2⅓

INSTRUMENTS

Speedometer at 30 m.p.h.	3% fast
Speedometer at 60 m.p.h.	accurate
Speedometer at 90 m.p.h.	3% fast
Distance recorder	2½% fast

WEIGHT

Kerb weight (unladen, but with oil, coolant and fuel for approximately 50 miles)	13 cwt.
Front/rear distribution of kerb weight	63/37
Weight laden as tested	16¾ cwt.

Specification

Engine

Cylinders	4
Bore	70·6 mm.
Stroke	68·26 mm.
Cubic capacity	1,071 c.c.
Piston area	24·3 sq. in.
Valves	Overhead (pushrod)
Compression ratio	9/1
Carburetters	Twin S.U. Type HS2
Fuel pump	S.U. electric
Ignition timing control	Centrifugal and vacuum
Oil filter	External full flow
Maximum power (net)	70 b.h.p.
at	6,000 r.p.m.
Maximum torque (net)	62 lb. ft.
at	4,500 r.p.m.
Piston speed at maximum b.h.p.	2,690 ft./min.

Transmission

Clutch	BMC, 7¼ in. single dry plate
Top gear	3·765
3rd gear	5·109
2nd gear	7·213
1st gear	12·047
Reverse	12.047
Final drive	Helical gears from transverse gearbox
Top gear m.p.h. at 1,000 r.p.m.	14·7
Top gear m.p.h. at 1,000 ft./min. piston speed	32·8

Chassis

Brakes: Lockheed hydraulic; disc front, drum rear, with Hydrovac servo

Brake dimensions:
Front discs .. 7½ in. dia.
Rear drums, 7 in. dia. × 1¼ in. wide

Friction areas: 51 sq. in. of friction lining (17·3 front, 33·7 rear) operating on 179 sq. in. swept area of discs and drums

Suspension:
Front : Independent by rubber springs and transverse wishbones. Rear : Independent by rubber springs and trailing arms.

Shock absorbers:
Front: Telescopic hydraulic
Rear: Telescopic hydraulic

Steering gear: Rack and pinion

Tyres: Dunlop SP 145—10 (5·5—10) (Dunlop C41 optional)

Morris Cooper S

A FEW years ago the idea of any Morris or Austin being eligible for competition work as it stood would have seemed fantastic. Yet today the homely-looking Mini-Cooper is one of the most respected performers in saloon car racing and rallies.

Testing the Mini-Cooper in its 997 c.c. 55 b.h.p. road form when it first appeared in the autumn of 1961, *The Motor* called it "a wolf cub in sheep's clothing." The new S type competition version, capping that performance substantially, surely rates as a full-grown wolf, in potentialities if not size. Thanks to a 1,071 c.c. engine embodying many features of the B.M.C. Formula Junior racing engine and giving 15 more b.h.p. than the normal Mini-Cooper, the maximum speed is nearly 10 m.p.h. up at 94.5 m.p.h. and 0-50 m.p.h. acceleration is 9.0 sec. as against 11.8 sec. The performance, moreover, is controlled by remarkable braking power.

Although the engine in this state of tune makes itself both felt and heard, the car is completely tractable, giving as good a service for shopping on Saturday morning as when racing on the Silverstone club circuit in the afternoon or rallying the same night. At a price of under £700 inclusive of Purchase Tax, this is one of the most inexpensive, most versatile and most exhilarating road cars ever offered.

How it does it

THE S type engine retains the 68.26 mm. stroke of the standard A-series Mini engine, but the bore is enlarged from 62.43 mm. to 70.6 mm., by making the centres of Nos. 2 and 3 cylinders ¼ in. closer, and the outer pairs ¼ in. wider apart. Sturdier but lighter connecting rods have offset little ends, the main and big-end bearings are of indium-infused

Little innocent: only an 'S' above the nameplate, and the perforated wheels carrying Dunlop SP tyres betray the "wolf in sheep's clothing" character of B.M.C.'s latest high performance Mini variant.

In Brief

Price (as tested) £575 plus purchase tax £120 7s. 1d. equals £695 7s. 1d.		
Capacity		1,071 c.c.
Unladen kerb weight		13 cwt.
Acceleration:		
20-40 m.p.h. in top gear		10.4 sec.
0-50 m.p.h. through gears		9.0 sec.
Maximum top gear gradient		1 in 10.1
Maximum speed		94.5 m.p.h.
Overall fuel consumption		26.8 m.p.g.
Touring fuel consumption		39.5 m.p.g.
Gearing: 14.7 m.p.h. in top gear at 1,000 r.p.m.		

Instruments (*right*). The Mini-Cooper S has only three dials: a 120 m.p.h. speedometer with fuel gauge and mileage recorder, flanked by oil pressure and water temperature gauges. The new fresh air heater below the parcels shelf is now standard equipment on all Mini-Coopers.

Furnishings (*below*) include Vynide seating and carpeted floor, with no tunnel to restrict leg space in the back.

Stout heart (*below right*) of the S is this 1,071 c.c., 70 b.h.p., twin-carburetter version of B.M.C.'s famous A series pushrod o.h.v. engine. The brake servo can be seen on the left, below the heater duct.

Morris Cooper S

copper lead, and are now both of 2 in. diameter, and a stiffer nitrided crankshaft in EN 40B high-tensile steel is used. Forged steel rockers replace the normal fabricated type, while Stellite-tipped valves in Nimonic 80 are used, working in copper-nickel alloy guides.

As tested, the S had standard Mini-Cooper gear ratios and a 3.76 to 1 final drive, but a choice of closer ratios is available and also a 3.44/1 final drive. Second and third mainshaft gears run in needle rollers, and the 7½ in. diameter clutch has bonded linings to withstand the extra torque. All these, and other subtle changes are a direct inheritance of racing experience in the saloon and Formula Junior classes. The rest of the car, too, reflects such experience.

Lockheed front disc brakes are increased by ½ in. diameter to 7½ in., and by ⅛ in. in thickness to ⅜ in.; larger brake pads are also fitted, with a material gain in friction lining area, cooling area and thermal storage capacity, and a Hydrovac servo booster greatly reduces the required pedal pressure, giving extremely powerful braking. The well-known ADO 15 rubber suspension at front and rear is unchanged and Dunlop SP tyres are fitted as standard.

What it does

THE Mini-Cooper S is undoubtedly an extremely fast car under any road conditions. It can whip through heavy town traffic, traverse twisting, hilly roads at a fine gait, and cruise at 90 m.p.h. on the motorway. Starting is immediate, and choke is required for a short while only, after which the car is smooth, tractable and well-mannered at low speeds, quite suitable to go shopping in. Yet its speed and accelerative

qualities will seriously disrupt the complacency of many owners of sports cars and big-engined saloons, especially as it is coupled with equally quick handling qualities and Mini manoeuvrability.

On the test car, a mean maximum speed of 94.5 m.p.h., and a best one-way quarter mile with a favourable wind of 98.9 m.p.h. were achieved. These are remarkable figures for a 1,100 c.c. saloon car capable of comfortably seating four people, and the acceleration is equally striking. From a stand-still, 30 m.p.h. takes just 4 sec., 0-60 m.p.h. is achieved in 12.9 sec., and 0-80 m.p.h. in 25.2 sec.

The performance has to be paid for somehow. Seventy brake horsepower from a small engine of "cooking" origin means a high degree of tune and a busy power unit. The test car developed a tendency to stall in traffic, remedied temporarily by occasional blipping of the throttle to prevent the rather rich mixture "ganging up" in the inlet manifold and eventually by adjustment to the carburetters, but the throttle linkage allowed a very variable idling speed. Although flexible enough at low speeds, the engine gets progressively happier as the revs rise.

At the absolute maximum of 62 m.p.h. in second gear, and 84 m.p.h. in third, vibration was carried through the gear lever to an uncomfortable extent. Although the engine note is taut and purposeful on brief acquaintance, the regular motorway traveller might soon tire of it at high speed, were it not for other compensations. Regrettably, no rev counter is fitted as standard, and gearchange points at 30, 50 and 70, which can be well exceeded, are marked in yellow on the 120 m.p.h. speedometer.

Fuel consumption, too, is naturally heavier at 26.8 m.p.g. than the normal Mini-Cooper's 34.6 m.p.g., but the gentler driver is rewarded by a touring consumption as high as 39.5 m.p.g. Engine oil on the hard-used test car was burnt at the rate of 135 miles per pint, a consumption attributable partly to the fact that the chrome scraper rings do not bed in very quickly, partly to the use of a larger capacity oil pump and enlarged oilways to ensure adequate lubrication under the stress of competition, and partly to the continual use of very high r.p.m.

How it handles

WITH so much power under the throttle foot, good road-holding and steering are essential, and the Mini system copes admirably. Use of the front drive "throttle-on" technique enables corners to be taken at remarkable speed, and although it proved possible to spin the wheels on tight, fast turns, the car remained controllable at all times. The light, high-geared rack and pinion steering with its immediate "feel" and response contribute vitally to the excellent handling and controllability, as do the braced tread Dunlop SP tyres on wheels which have rims one inch wider than normal. This combination gives extremely high cornering power and a rapidity of response to the steering which can make cornering rather jerky until a driver strange to the car acquires the necessary lightness of touch. When he does he finds that cornering limits are so high, both in the wet and the dry, that they are difficult to explore; there is no tyre squeal and these special tyres, used for some time on B.M.C. rally cars, reduce considerably the drift angles on bends and, in consequence, reduce also the usual change in handling between "throttle on" and "throttle off" conditions which becomes almost imperceptible when one is trying extremely hard.

The ratios of the four-speed gearbox are excellent for all-

Extra luggage (*above*) can be carried on the open boot-lid.

The S on the lid (*below*) suggests the shock the Mini-Cooper S can administer to many drivers of larger, more powerful cars.

round road use, and the remote-control gear lever is easy to reach. There is no synchromesh on first gear, which is an occasional inconvenience with an engine thriving on high revs. On the comparatively new test car, gear lever movement was a little stiff, while shopping drivers may not relish the fairly heavy clutch pedal pressure.

The braking powers of the S are formidable, and in complete harmony with its performance. The disc-front, drum-rear combination allied to Hydrovac servo assistance gives extremely effective braking with only gentle pedal pressure, while repeated use brought no evidence of brake fade. The "velvet touch" proved particularly necessary in the wet when over-hard application brought considerable initial pull until the discs dried off. An emergency stop from about 70 m.p.h. on dry roads, however, pulled the car up virtually all-square, with the slightest hint of wheel lock on the nearside.

During the test the brakes developed an intermittent tendency to stick on momentarily after releasing the pedal, spoiling the cleanness of the run-down to corners in the lower gears. Apart from this, they were beyond reproach.

There is some pitch and quite a lot of bounce over bumpy surfaces, the penalty of firm springs and lack of overhang, and bumps also affect the cornering, causing some wallow and tending to throw the car off line, especially on a trailing throttle. Otherwise, rough-road handling is good, and the "van" angle of the Mini's steering column provides excellent controllability and good all-round vision as well, all of which adds to the eminent safety of the Mini-Cooper.

Furnishings and Layout

THE front seats in this B.M.C. projectile are comfortable but upright in typical Mini fashion, but taller drivers find the legroom inadequate, even with the seat set right back. Cockpit fittings are fairly sparse. Set centrally in the dash is the oval instrument panel, containing a speedometer with fuel gauge inset, oil pressure gauge on the right, and water temperature gauge on the left. An impromptu running-out of petrol accentuated the inadequacy of a mere 5½-gallon tank; the optional extra tank of the same capacity will probably be demanded by most road and all rally users.

Below the dash, left to right, are the heater, wiper, ignition, headlight and choke controls, with lower down a kind of "gear gate" control for the very effective fresh-air heater system. All these are too far from the driver, especially if he is using safety belts. There is adequate floor space in front, as is usual with Minis, capacious parcel shelves, and very useful side boxes at the base of the two doors. The wipers are non-self-parking.

The headlights, into which are embodied the parking lights, give an excellent beam for night driving, but a "flasher" on the steering column would be very useful for warning the drivers of fast cars who are unused to seeing a Mini pushing hard in the mirror. The childish bleep of the horn, too, is utterly inadequate for a fast car and the washer button is not easy to reach. For the size of the car, the carpeted boot is spacious, while extra luggage can be carried on the opened lid. The spare wheel lives below the boot floor.

In all, the Mini-Cooper S is a car of delightful Jekyll and Hyde character, with astonishing performance concealed within its unpretentious Mini-Minor skin. More than just a "two-upmanship" car, it has a truly formidable competition potential at a very moderate price.

═══ Coachwork and Equipment ═══

Starting handle	None	Sun visors	2
Battery mounting	In boot	Instruments: Speedometer with fuel gauge and	
Jack	Side lifting	mileage recorder. Oil gauge. Water temperature gauge.	
Jacking points	One each side on body sill	Warning lights .. Ignition, headlamp, main beam	
Standard tool kit: Jack, wheel brace and jack handle, plug spanner, tommy bar.		Locks:	
Exterior lights .. Headlamps, side lamps, tail lamps		With ignition key .. Doors and trunk lid	
Number of electrical fuses	2	Glove lockers .. None	
Direction indicators .. Front and rear flashers		Map pockets .. One in each door, two behind	
Windscreen wipers .. Electric twin-blade, single speed		Parcel shelves .. One under facia and one behind rear seat	
Windscreen washers .. Wingard twin-jet		Ashtrays .. On top of facia, and in each companion box	

Cigar lighters ..	None
Interior lights ..	1
Interior heater .. Standard. Fresh air type	
Car radio .. Radiomobile (extra)	
Extras available: Britax safety belts (anchorages provided). Extra fuel tank (5½ gals.). Oil cooler. Sump guard.	
Upholstery material: Vynide seats. P.V.C. leather-cloth roof lining.	
Floor covering ..	Carpet
Exterior colours standardized ..	6
Alternative body styles ..	None

═══ Maintenance ═══

Sump and transmission .. 8 pints Multi-grade, plus 1 pint for filter		Contact breaker gap .. .015 in.	
		Sparking plug type .. Champion N5 long reach	
Steering gear lubricant .. Hypoid SAE 140		Sparking plug gap .. .025 in.	
Cooling system capacity: 5½ pints, plus 1 pint for heater (2 drain taps).		Valve timing: Inlet opens 5° b.t.d.c. and closes 45° a.b.d.c. Exhaust opens 51° b.b.d.c. and closes 21° a.t.d.c.	
Chassis lubrication: By grease gun every 3,000 miles to 8 points.		Tappet clearances (cold): Inlet .015 in. at valve. Exhaust .015 in. at valve.	
Ignition timing .. 3° b.t.d.c.		Front wheel toe-out .. 1/16 in.	

Camber angle .. 1° min.; 3° max.	
Castor angle .. 3° nominal	
Steering swivel pin inclination .. 9¼°	
Tyre pressures: Front, 24 lb. Rear, 22 lb.	
Brake fluid .. Lockheed 328	
Battery type and capacity .. 12 volt Lucas, 43 amp.-hr. at 20 hr. rate	

Foley and friend; understeer onto the Causeway.

THE SHATTERING

TODAY perhaps the most interesting and spectacular racing to be seen on circuits throughout the world is sedan or touring car racing. Admittedly Grand Prix racing and sports car racing is infinitely faster, but the development of these types of cars has reached such a high point that for the spectator they no longer have that spectacular look of the cars of the fifties; all tail out or running off the road under full lock.' Today the sedan has replaced them as sheer spectacle.

Most people go to motor races today not really because they like the bark of Climax motors or revel in identifying the prostrate figure busily sorting cog from cog in the belly-pan of a Brabham. They go to races to watch all this on a broad canvas but they also attend to see shoeboxes sliding sideways, Cortinas endeavoring to devour 4.2 Jaguars and their car beating their neighbor's.

At first manufacturers passed off this form of the sport as a little bit of spoofery for the eccentrics. But gradually over the last few years they have taken more and more notice. Previously

It may not be as fast as a 4.2 Jaguar but it will sure give them something to worry about. By CHRIS BECK

most public relations departments relied on trials and rallies to prove the worth of a particular vehicle to the public. But today they divert more money and time to sedan car racing. Consequently the demand for a special production car intended solely for racing has developed. Usually the car remains basically the same as its mass production brother, but is fitted with different engines, transmissions and modified suspensions. The Morris Cooper 'S' is one of these.

Australia has only seen a little of the Morris Cooper 'S' — which will be called in this article

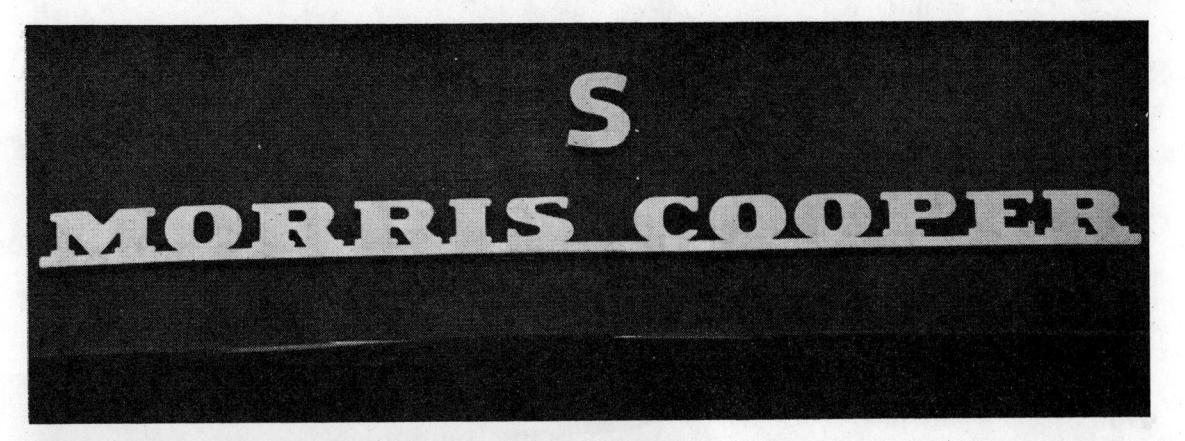

Only a small 'S' above the badge front and rear and the drilled wheels distinguish the car from a normal Morris Cooper. Hard to see at 90 mph.

The trumpets of the Weber carburettor hide behind the metal panel which covers the hole where the speedometer normally is. The speedometer has been removed to the left hand parcel shelf, where driver can't see it.

COOPER'S'

the S-type — for at the present only five have been prepared for competition.

One of them is the car for this month's track test.

When we heard that P. and R. Williams, a large Sydney BMC distributor, had finished preparing an 'S' for one its sales executives, Brian Foley, to race, we reached for the phone. Sales chief Tony Donkin gave an immediate OK, but asked us to hold off until the car had run in the 50 mile AJC Trophy at Warwick Farm last December. In the race it won its class and established a new lap record for under 1300 cc. After the meeting the car was stripped down in preparation for the International series. When it was finally nailed together again we met it at the test day at Warwick Farm — the short circuit.

From the front the squat, pugnacious little car resembles a chunky-shouldered bulldog. The extensively lowered front suspension — attained by cutting the rubber trumpets — has brought the body down almost 2.5 ins. The same method has been used at the rear, but to lower it only

2 ins. Double-acting adjustable Koni shock absorbers replace the original units at the front while the rear carries standard dampers with slightly heavier compression settings. Wide-rim 4.5 in wheels are shod with Dunlop R6 racing booties; the first ten-inchers to arrive here and just the same as McLaren uses on the open wheelers, tread, mixture and all. At the front the wheels bolt directly to the hub but at the rear there is an extension from the brake drum onto which the wheels bolt. Half a degree of positive camber and a .125 in. toe-in finish the suspension arrangement.

These suspension modifications — completely useless for road work — go hand in hand with the alterations made to the power unit. To make full use of the race suspension it is obvious that the car has to have a motor producing a substantial amount of power. And power it has got. There are few engines, including those used in

Formula Junior machinery, that could match this A-series powerhouse. So far the bhp output has not been made public, but we can say that the motor is producing well over 100 bhp per litre.

The S-type uses the substantially oversquare 1071 cc engine developed in England by BMC and John Cooper for the latter's FJ cars. Similar motors in Britain have been dynamometer-recorded approaching 10,000 rpm.

When P. and Rs received the car it was stock standard and according to the factory catalogue produced 70 bhp. Peter Molloy, custodian of the company's Austin Healey Sprite (track tested January '64) was told to make the S-type go as fast as humanly possible. Three weeks later, just before the AJC Trophy, the car was ready to run. Then Molloy was obtaining dynamometer readings of 89 bhp.

After the race he rebuilt the car again. First came the cylinder head, which had already been worked almost to FJ standard by the factory. More modifications and reshaping of the combustion were done and the compression ratio was raised from 9.5 to 1 to 13.5 to 1 by the fitting of flat-top pistons instead of the concave units normally found in S-types. The standard 1 11/32 in. inlet valves and 1 1/8th in. exhaust valves were highly finished and special single valve springs replaced the normal dual contra-coils. No modification was needed to the bronze guides and standard collets were used in conjunction with lightweight aircraft alloy valve caps which were specially machined. The rockers on the tappet gear are fitted, like all S-types, with needle rollers and the tension springs between the rockers have

Foley and Molloy discuss the procedure before the running in period. Note how low to the ground the car is at the front.

been replaced by tubular bronze spacers to cut down friction at high rpm. To stop gasket-blowing between three and four cylinders, a common fault with S-types racing overseas, hand-made copper asbestos gaskets are used.

Built differently from the normal Cooper in that the centre main bearing is not as wide and the two end mains are wider, the bottom end remains completely standard except for balancing. All bearings on the big ends and mains are made of lead indium for a higher wear rate. Weighing only 9 lb, half the weight of a normal Morris Cooper unit, the flywheel fitted to the Cooper S-type is lightweight alloy. A standard clutch with beefed-up spring pressures completes the assembly and the lot has been completely balanced.

Unlike normal S-types, which are fed by twin SU carburettors Molloy has fitted a double choke DCOE 40 Weber carburettor. He says the Weber is easier to tune and produces a great deal more power right throughout the rev range. To fit the side-draft Weber, he had to remove the speedometer housing to get room for the two throats of the carburettor.

Continued on page 30

The 4.5 inch rims put a lot of the R6 Dunlop racing tyre on the track. Both tyres front and rear protrude outside the bodywork. The track has been widened by almost an inch.

The motive force; this little 1071 cc motor is developing more than 100 bhp/litre, enough power to beat many of our better FJ cars.

THE SHATTERING COOPER 'S'

Continued from page 28

Molloy had a special three-branch extractor exhaust made to pass the gases quickly from the cylinders into a 1.5 in. diameter pipe, to which is connected a small Lukey echo chamber-type muffler designed to take a little of the bite out of the noise of the hard working little engine.

The baulk ring synchromesh in the gearbox runs on needle rollers instead of plain metal bushes as in the normal Morris Cooper and generally is a more robust box. Things have been strengthened to take the higher power output delivered by the S-type motor. When the car arrived from England it was fitted with a high 3.4 to 1 touring final drive so Molloy replaced this with a 4.1 unit. The remote shift mechanism remains as standard. The sump, oil galleries and external oil cooler, fitted beneath the front bumper bar, hold a total of 9 pints. Molloy modified the oil pump by enlarging and polishing the ports for a quicker and smoother flow. The water radiator and cooling system is standard.

At present the gear ratios, although close enough for road use, are not the right ones for racing. Molloy is at present awaiting high first and second gears from the UK.

Inside, the car is not as well finished as the Australian Morris Cooper. The upholstery looks cheap and the general appearance is of lesser quality.

At the Farm test day we found that the car had not been run since the AJC Trophy and that Peter had fitted new bearings so the car had to tolerate a tedious running-in procedure. Brian took the car out for 10 laps at 3000 rpm and when he had completed this he added 500 rpm every couple of laps or so until by the 30th lap he felt the bearings had been bedded in and full power and revs could be used. Every three or four laps he would come in to check the pressure of the D6 tyres. Generally 50 *psi* front and 45 *psi* rear seemed to be the best combination and Molloy would at the same time take a peek at the Bosch racing plugs.

After Brian had done 10 or twelve quick laps and recorded 47.9 seconds for the 9/10 mile circuit he brought the car in and handed it over.

Starting the little motor is no trouble; it coughs a couple of times and then catches to burst into life with a subdued bark. The gearbox has the same pattern as the normal Morris Cooper and is as easy to operate. Applying the power and dropping the clutch in first gear provokes far too much wheelspin; you hold this gear to 6500 for a quick change to second, and again the wheels spin madly; and, again, upon the change to third strips of rubber are left on the road.

After a couple of familiarisation laps I came in and then after a few words took the car out again. This time I had really got the feel of it. As well as the phenomenal acceleration—which is most un-Mini-like — I began to appreciate the car's magnificent handling qualities. One did not seem to be able to go into any corner fast enough and whenever the power was called in in third the wheels would momentarily break adhesion. A little wheelspin could also be induced on a change to fourth.

By racing sedan standards the inside of the car was relatively quiet and there seemed to be no flexing. Without any trouble whatsoever a lower limit of 6000 rpm could be maintained around the entire short circuit.

The P. and R. Williams' S-type firmly illustrates that the refinement in racing sedans today virtually make it impossible for the family man to drive his car on the road during the week and campaign it on weekends.

Indeed, the racing sedan car takes as much if not more development than the single-seater F2 and FJ, or even an ANF car. I sincerely hope that through this development sedan racing does not lose its sheer spectacle.   #

PATRICK McNALLY TESTS THE SPEEDWELL CLUBMAN MINI-COOPER

I HAVE driven more than a few modified Minis in past years and they all have had one feature in common: when driven over 100 m.p.h. they felt as if they would burst at any moment! It was very pleasing, therefore, to find the Speedwell Clubman Mini-Cooper was turning over at a mere 6,150 r.p.m. at 100 m.p.h.

Speedwell's have set out to modify the Mini-Cooper engine to such a degree that the car is now capable of employing a 3.44 final drive without spoiling the rapid acceleration. This they have done with more than a fair measure of success.

The car I tested was an ordinary Mini-Cooper which had been bored to 1,152 c.c. and fitted with special flat-top pistons. The cast iron cylinder head had been discarded in favour of an alloy head of Speedwell's own design. This has iron valve seat inserts and is machined to a high standard of engineering, employs an intriguing combustion chamber shape and is highly polished. The use of aluminium permits a higher compression ratio to be used without any "running on" troubles—and there are, of course, the other advantages of using aluminium.

A compression ratio of 11:1 is used. The valve gear, which includes special valve springs and high lift rockers, are all of Speedwell design. The camshaft, Speedwell call it a C.S.6 (which is approximately the same as the B.M.C. A.E.A. 544), gives good torque over 3,500 r.p.m. The whole unit is balanced and the flywheel is considerably lightened. 1½ in. SUs supply the mixture through a standard induction manifold which has been polished and matched to the ports. The main bearings are located in beefed-up big ends, although the standard con-rods are retained.

Items like the oil pump and the distributor receive special attention, the distributor advance being matched to the camshaft and the oil pump being replaced by a new unit of Hobourn-Eaton manufacture.

The engine produces 82 b.h.p. at 6,800 r.p.m. and 70 lb.-ft. at 4,500 r.p.m. The exceedingly good torque figure is where Speedwell score, for it is this feature that permits the high axle to be used without killing the acceleration. The test car, rather than lightened, was completely trimmed and had some useful optional equipment. This included a Speedwell rev. counter and a combined oil pressure and water temperature gauge. As far as I am concerned, rather than being optional equipment, they would appear vital necessities to a car of this calibre.

The car proved extremely lively on the road, and the high axle proved to be a great advantage when coupled to the standard ratio Cooper baulk-ring gearbox. 63 m.p.h. was possible in second and 89 m.p.h. could be found in third, presuming one went up to the red —this started and finished at 7,500 r.p.m.

The acceleration would have been good for a car with a 4.1 final drive; as it was a 3.4:1 it was truly excellent. If speed was the essence (as it nearly always was) 60 can be reached in 8.4 secs. Without causing comment from an elderly passenger, who no doubt

would be admiring the scenery, the same speed could be reached in 9 secs.

The rest of the figures are all of this order with 30 m.p.h. coming up in 2.85 secs. and 50 in 6.45 secs., 80 taking 15.5 secs. The quarter mile was covered in a shattering time of 16.9 secs.; the worst figure recorded for this distance was 17.1 secs. on a slightly damp road.

Maximum speed caused some head-scratching until it was realized that a strong wind was causing the two-way maximum to vary by no less than 8 per cent. Under better conditions a mean speed of 109 m.p.h. was clocked. With the following wind we earlier recorded a one-way time of 113.9 m.p.h.—no doubt the Mini hoisted a spinnaker when the driver was concentrating on the matter in hand!

With such small wheels there is always the danger of excessive heat in the tyre department. So for the purposes of acceleration tests 35 p.s.i. was used in the front tyres, with the exception of maximum speed runs, when a further 15 p.s.i. was added.

On the debit side it must be said that, although the car was very tractable as long as the revs were kept over 3,000, below this the engine was a little on the rough side. Indeed, idling at 1,000 r.p.m., the combination of under-bonnet noises made the engine sound like the proverbial bag of nails. However, this is a small price to' pay for 100 m.p.h. motoring and 8.4 secs to 60 m.p.h.

After driving from one side of London to the other, the plugs sooted

up, but once the car was on the open road they cleared themselves without any trouble. The test was conducted without the endless changing of plugs which tends to dampen the spirits of even the most enthusiastic.

On one occasion (when I was concentrating on popsies and not petrol pumps), I succeeded in having the car filled with premium fuel—not 100 octane. The car behaved exceptionally well. The performance was not noticeably affected, the only tell-tale being that the car ran on. On checking my bills I discovered the error, and a plug check revealed a slightly weaker mixture (not surprisingly), otherwise there were no other outward signs of the combination of 11:1 c.r. and 94 octane fuel.

Restall seats were fittted and these

were extremely comfortable, affording support in all the right places. These seats are fully reclining and can be manoeuvred so as to form a bed. Perhaps unfortunately, I had no cause to spend a whole night in the car, so I had to take the Speedwell's word when they said "they're just the job for a little rest"—or something like that anyway.

To continue: the exhaust includes a fabricated manifold, and a twin silencer was adequate, though perhaps a little on the noisy side. This feature might be considered an advantage by some, but personally I find a car can always be driven that little bit quicker on the road when it's quiet.

The suspension was standard, save for the fitting of an anti-roll bar to the rear. This dampened the car's under-steering characteristics a little. For maximum benefit from the roll bar a 10 lb. differential between front and rear tyres (35 p.s.i. front, 45 p.s.i. rear) gave more exciting handling—not bags of oversteer, but adequate. The Kenlowe electric fan, which was thermostatically controlled, kept the water temperature down and caused not a moment's worry. Oil pressure remained steady and high, consistently over 60 lb. per square inch.

Remarkable features were the low oil and petrol consumptions. Petrol was consumed at the rate of a gallon for 26 miles at its worst and up to 28 m.p.g. on longer journeys. Oil consumption, the bugbear of bored out Mini-Coopers, was really good and only four pints were added in the thousand miles covered during the test. Garage space being limited, I had to leave the Mini-Cooper under the stars on two occasions. Nevertheless, it always started on the first touch of the button.

Big-bore Mini-Coopers are the most

THE CLUBMAN has had the usual cast iron head replaced by a Speedwell alloy version; also 1½ in. SUs are added.

exciting cars to drive if only people who are about to be vanquished by the little bombs could let one pass without all the anti-Cooper manoeuvres!

Performance. Speeds in the gears: 1st, 38 m.p.h.; 2nd, 63 m.p.h.; 3rd, 89 m.p.h. **Standing ¼-mile** 16.9 secs. 0–30 m.p.h., 2.85 secs.; 0–50 m.p.h., 6.4 secs.; 0–60 m.p.h., 8.4 secs.; 0.80 m.p.h., 15.5 secs. **Maximum speed:** 109 m.p.h.
Engine conversion: £200.
Cost of complete car: £800.
Optional extras fitted to the Test Car Included: Speedwell Exhaust Manifold, £8 15s.; 3.44:1 High Final Drive, £5; Restall de Luxe Seats, £25 each; S.P.C. Rear Anti-Roll Bar, £7; Headlamp Flasher, 18s. 6d.; Wooden Dashpanel, £3 7s. 6d.; S.P.C. Combined Gauge, £5 10s.; S.P.C. Electronic Tacho, £17 10s.; Brake Servo, £13 10s.; ·S.P.C. Wooden Steering Wheel, £11 19s. 6d.; Oil Cooler, £13 10s.; Kenlowe Fan, £12 10s.

Whitsun sport

S day at Mallory

Waiting to pounce? No. 2 is the Banks/Rhodes Mini-Cooper which won the race — 4 is the Weber/Cabral privately entered one which kept ahead for so long—until half distance.

Works Mini-Coopers take first and second in 'Motor' Three Hours race and vanquish the Continentals

UNLIKE last year, when Saab and B.M.W. fought for the lead in the *Motor* Three Hours saloon car race at Mallory Park to the last lap, this year's race was a walkover for the 997 c.c. Mini-Cooper S works cars of Warwick Banks/John Rhodes and John Fitzpatrick/Julien Vernaeve. Another S, driven by John Thurston and John Terry, was third, but the

Saab-B.M.W. battle was repeated, with the Swedish two-strokes narrowly taking 4th and 5th places ahead of the German flat-twins, both winning their classes.

The weather was superbly warm and sunny on Whit Saturday afternoon when the field of 18 up to 1,000 c.c. cars lined up on the grid for the 3 p.m. start. Positions were decided, not by practice times

Winners' reward —the "Motor" Trophy being handed over by Harold Hastings (the one with the glasses) to John Rhodes and Warwick Banks.

but on engine capacities, the "1000s" to the front, with balloting to decide exact positions. John Fitzpatrick had, however, lapped in 1 min. 3·6 sec. in practice with one works Mini-Cooper S, while Warwick Banks and the Downton-prepared S of Tommy Weber/Mario Cabral both did 1 min. 3·8 sec.

With Swedish, German, Dutch, Belgian, Swiss and Portuguese drivers and Swedish, German, and Italian as well as British cars taking part, this "National" Three-Hours was far more International in character than many a Formula 1 race. Several drivers had come from racing at Zolder, Belgium, the previous week, in quest of points for the up to 1,000 c.c. classes of the 1964 European Touring Car Championship, the Mallory race constituting the second qualifying round in this contest.

Practice brought its troubles. The German D.K.W. two-stroke expert Dieter Mantzel blew up an engine, and was still busy fitting a spare during the last hour before starting time; his method of making the work easier by heaving the entire car up and leaning it bodily at a steep angle on a spare wheel greatly intrigued paddock spectators! Other late toilers were the Dutchmen Swart and Keller on their Fiat-Abarth, while the Else/Addicott D.K.W. 800S was unfit and non-started.

To have a young lady to hold up the " Mins. to go " boards was a good idea by the organizers, the B.R.S.C.C., to ensure maximum attention. Soon after she held up the "2 Mins.", engines were started, producing a rare assortment of noises, the healthy rasp of Mini-Coopers mixing with the deep note of flat-twin B.M.W.s, the wroom-doom-doom of D.K.W.s and the high whoop of the two red Saabs.

With four Minis in the front row, four Minis not unnaturally led into the first corner at flag-fall, but Mantzel in his raucous D.K.W. checked these monopolist tendencies by taking Vernaeve's fourth place on the second round. After another stormy lap, however, the German stopped at Shaw's Corner and wrestled long and uncomfortably with a practically red-hot starter motor which had remained engaged with the engine. He finally succeeded in disengaging it, and restarted by running backwards down the slope. This contravened rule 7 of the supplementary regulations entailing disqualification, but Mantzel resumed racing with spasmodic speed and intermittent stops, one

a fine steamy overheated one, before yielding to the inevitable and giving up.

With Mantzel went no small part of the race interest, for the Saabs and B.M.W.s were not fast enough to disrupt the Mini S monopoly, and only the impudence of Tommy Weber's private S in leading the works cars kept boredom at bay. For lap after lap he and John Rhodes flogged round nose to tail, neither giving an inch, and steadily leaving Julien Vernaeve's second works Mini-Cooper and the rest behind.

After covering 22 of Mallory's short sinuous laps the leading pair were passing the tailenders for the second time, and by the first hour Weber had completed 54 laps, averaging 75·42 m.p.h. and still led by the proverbial cat's whisker from Rhodes, Vernaeve, Thurston and Grahser's 700 B.M.W. which had somehow got ahead of the noisy Saabs. The W. E. Allen/P. Kelly Mini-Cooper S was in trouble with a wheel bearing, and the Anstead/Gunther Fiat-Abarth was another frequent pits visitor, first with

PROVISIONAL RESULTS

"Motor" National 3-hour Race (Saloons)
Overall placings: 1, W. Banks/J. Rhodes (Austin Mini-Cooper S), 165 laps, 74·51 m.p.h.; 2, J. Fitzpatrick/J. Vernaeve (Morris Mini-Cooper S), 164 laps; 3, J. C. Thurston/J. V. Terry (Morris Mini-Cooper S), 158 laps; 4, B. Rothstein/B. Johansson (Saab 96), 157 laps; 5, G. Karlsson/S. Johansson (Saab 96), 157 laps; 6, W. Schneider/H. Hahne (B.M.W. 700 CS), 157 laps; 7, J. Grahser/P. Marx (B.M.W. 700 CS), 155 laps; 8, T. Weber/M. Cabral (Morris Mini-Cooper S), 148 laps; 9, S. Thynne/C. Stancomb (Morris Mini-Cooper S), 146 laps; 10, H. E. B. Mayes/J. R. Aley (D.K.W. Junior), 144 laps; 11, P. K. Wilson/J. F. Harris (B.M.W. 700 CS), 139 laps; 12, W. D. Kelly/P. Wicks (Morris Mini-Cooper S), 128 laps; 13, H. Gilges/H. Vogel (B.M.W. 700 CS), 129 laps; 14, E. H. Swart/P. A. Keller (Fiat-Abarth 850 TC), 127 laps.
Class Placings. Class A, 851-1,000 c.c.: 1, W. Banks/J. Rhodes (Mini-Cooper S); 2, J. Fitzpatrick/J. Vernaeve (Mini-Cooper S); 3, J. C. Thurston/J. V. Terry (Mini-Cooper S). **Class B, 701-850 c.c.:** 1, B. Rothstein/B. Johansson (Saab), 70·81 m.p.h.; 2, G. Karlsson/S. Johansson (Saab); 3, H. E. B. Mayes/J. R. Aley (D.K.W.). **Up to 700 c.c.:** 1, W. Schneider/H. Hahne (B.M.W.), 70·73 m.p.h.; 2, J. Grahser/P. Marx (B.M.W.); 3, P. K. Wilson/J. F. Harris (B.M.W.). **Team Award:** Cooper Car Co. (W. Banks/J. Rhodes, J. Fitzpatrick/J. Vernaeve and J. C. Thurston/J. V. Terry).

Deutcher dice: the Mayes D.K.W. Junior (No. 26) and the Grahser B.M.W The 'Deek' finished in 10th place and the 'Bee Emm' made it in seventh.

misfiring, then with starter motor ailments.

Interest mounted with the half-distance pit stops for fuel, tyres and driver changes. These also settled the destiny of the race, for Weber's hard-won lead was dissipated in a pit stop lasting 2 min. 49 sec. in which both front tyres were changed, using a narrow, precarious jack, and the Portuguese driver Mario Cabral took over the driving. In contrast, the works Mini-Cooper stops, presided over by Ken Tyrrell, occupied less than a minute, with only the nearside front wheels, their tyres virtually treadless, being changed, and the pit doing their own refuelling, thank you, rather than use the "filling station" installed beyond the pits and used by most competitors.

Saab pitwork, accompanied by much shouting in intriguing Swedish, was also expert, but several other teams threw away many vital seconds which could

"Four Minis in the front row ... four Minis not unnaturally led into the first corner at flag-fall. ..."

S day at Mallory

Continued

never be retrieved on the short, tricky circuit. By the time the flurry and horn-blowing at the pits were over and the traffic jams of Saabs, Minis and B.M.W.s were reforming on the back straight, the situation looked highly favourable for the works Mini-Coopers following that gift of two laps from the Weber car. After 100 laps, with 1¼ hours to go, their pit gave a satisfied " P1+10 " signal to Warwick Banks in car No. 4, signifying Place 1, 10 sec. lead, and "P2 4—8" (2nd, 8 sec. behind No. 4) to Fitzpatrick.

Not that Cabral accepted the situation without fighting. Coming up with the leading Mini, he passed and set out to try and retrieve those lost laps. The gap opened visibly; at 5 p.m., when Banks had covered 110 laps, there were 5 sec. between his and Cabral's car; another lap and it was 5·8, and the next time 7·6.

The Portuguese was pulling out over a second per lap, but with under an hour to go his task looked hopeless. Ken Tyrrell knew it and held his car's pace, but Cabral pressed on valiantly, until Banks's car was out of his mirror. 13 sec., 15, 18, 19, 21, the gap widened until, with 40 mins. to go, Cabral had 65 sec. to make up. It was too much, although hopes in the Weber camp arose on seeing " HOW FUEL " signs going out to Banks and Fitzpatrick. Alas, both drivers responded with cheerful thumbs up, and at 5.41 Cabral, possibly tiring, dropped 2·4 of his hard won seconds.

Two laps later he didn't come round at all, a sudden focusing of attention and binoculars on Gerard's Bend signifying the end of his efforts. The nearside front hub of the Mini-Cooper had broken, and the car left the course, hit the bank, bounced off and hit it again. Poor Cabral suffered head cuts and was taken by motorboat across the lake to the ambulance and then to hospital—alongside a severe case of sunstroke—in England!

Meantime, troubles had struck elsewhere. The Allen/Kelly Mini-Cooper S had finally retired, the Thynne/Stancomb and Kelly/Wicks S Minis both had plug trouble, the Gilges/Vogel B.M.W. broke a wheel, the Wilson/Harris B.M.W. stopped on the circuit, its gearbox minus oil and drain plug, and the Dutchmen Swart and Keller had dynamo difficulties on their Fiat-Abarth.

Banks and Fitzpatrick romped on unchallenged, and only a late effort by Gosta Karlsson's Saab to pass the Schneider/Hahne B.M.W. enlivened the final quarter of an hour, while the Gilges/Vogel B.M.W. re-wheeled, stopped again at Gerard's, bringing out the yellow flags with 5 minutes to go. Slowly the German car got going again, to complete the lap while Warwick Banks and John Fitzpatrick took up victory formation and crossed the line nose to tail to take the chequered flag.

The pit-stop that cost the Weber Mini more than two laps — Cabral (nearest camera) did his best to make it up but mechanical troubles took a hand.

The Thurston/Terry S was third, 7 laps behind, followed by last year's race winner Bjorn Rothstein/B. Johansson, class winners again this year in their Saab. Then came Karlsson's sister car, 700 c.c. class winners Schneider/Hahne in their B.M.W., and Grahser, while everyone was pleased that, despite its mishaps, the Weber/Cabral Mini-Cooper S was classified eighth overall and fourth in the 851-1,000 c.c. class.

— and on Sunday

THE Sunday programme at Mallory Park embraced five races, four of which were won by Scotsmen. Jackie Stewart scored yet another Formula 3 victory in Race 1, for the *Express and Star* Trophy, he and his team mate Warwick Banks scorching round in the Tyrrell works Cooper-B.M.C.s, sharing the fastest lap and finishing a length apart, well ahead of David Baker's M.R.P. Lola and Jim Russell pupil Melvin Long in a Lotus.

Stewart then climbed into one of the Chequered Flag Lotus Elans for the 10-lap race for up to 1,600 c.c. GT cars. With Peter Arundell, Mike Spence and Mike Beckwith also in Elans, it looked like a tight race, but Stewart made a brilliant start to head Arundell off into Gerard's; Arundell pressed hard but could not catch the flying Scot, who pulled out a 2·2 sec. lead and also set a new class lap record at 88·04 m.p.h. Brian Bennett had several nasty moments when his Turner bonnet flew open at the first corner, R. Nash's Marcos hitting him amidships and both retiring.

Next came the " big bangers' " turn, the Guards Trophy 20-lapper for over 1,150 c.c. sports cars. Three of the 4·7-litre Lotus-Ford Thirties entered were non-starters, but with Jim Clark present in the sole car completed so far interest remained high, particularly as he had

Continued

Clark blasts Indy record

INDIANAPOLIS, SATURDAY: Jim Clark (Lotus/Ford) today shattered the track record in practice at 159·377 m.p.h.—7·53 m.p.h. better than the old figure. He also set a four-lap average record at 158·838 (previous best 151·153).

The first 18 qualifiers broke the previous lap record 18 times and of the 21 cars which have so far qualified, 13 are rear-engined and six of them powered by the new Ford engine, fully described in next week's *Motor*.

Three hours after his record-setting, Clark flew home to compete at Mallory and Crystal Palace.

Dan Gurney qualified at 154·487 m.p.h. in the other Lotus Team car. Qualification runs continue next Saturday and Sunday and among those still in the queue are Masten Gregory, Mickey Thompson and Jack Brabham, who was waiting to come out when the track closed for the day. Jack now has to choose between missing Indianapolis qualification or the Dutch GP next week-end.

Banks and Stewart in the Tyrell Coopers lead off in the Formula Three event.

S day at Mallory

Continued

taken the brute round the 1·35-mile circuit in 53·2 sec. (91·35 m.p.h.) shortly after reaching Mallory from Indianapolis. News of his fantastic 159 m.p.h. lap there gained him a big cheer.

Clark's main opposition, Roy Pierpoint's 4·7 Ford-engined Attila, leapt into the lead while the Lotus's rear wheels became a smoky blur, but once moving the World Champion was soon on the Attila's tail. Lap 3 and he was past, after which it was all over bar the spectators' gasps at the big Lotus's acceleration out of the hairpin. A rather smoky Attila came home 20·6 sec. behind, while third —and first in the up to 2-litre class— was Roger Nathan's Repco-Brabham Climax, which was followed by Julian Sutton's Lotus.

Then came the big race of the day, the Grovewood Trophy Formula 2 event over 30 laps. Jim Clark had one works Lotus, Arundell the other, and there were four other Lotuses, three Lolas, three

Coopers and four Brabhams, a Merlyn and an Alexis, every one of them powered by Cosworth-Ford. Sensation of practice was young Jochem Rindt of Vienna, who made the fastest practice lap on a circuit new to him in 52·4 sec. with his blue and white Brabham—2 sec. quicker than Jim Clark, Brian Hart and Denny Hulme.

Rindt's pole start was ruined by a dud clutch, and while Hart, Clark and Arundell shot their Lotuses to the front, the Austrian was 13th. Clark took the lead from Hart on lap 3, Arundell passed Hart on lap 4, and Gardner (Brabham) did likewise on lap 8. Meantime, Bill Bradley's Lola had spun at Gerard's on the third round, hitting an all too substantial marshal's post, and involving his team mate Dick Attwood and Kurt Ahrens' Cooper, all three retiring.

Rindt by then was 10th and clutchless. but despite this he moved up to 9th on lap 5, to 8th on lap 10, to 7th on lap 18, (when Hart's Lotus retired with misfiring later traced to a loose plug lead), to 6th on lap 19, to 5th on lap 20, to 4th on lap 27, and to third behind the Lotuses of Clark and Arundell on lap 28—a remarkable drive by a promising driver.

PROVISIONAL RESULTS

" Express and Star " Trophy (Formula 3, 15 laps)
1. J. Y. Stewart (Cooper-B.M.C.), 14 min. 30·0 sec., 83·80 m.p.h.; 2, W. Banks (Cooper-B.M.C.), 14 min. 30·4 sec.; 3, D. Baker (Lola-B.M.C.), 14 min. 55·0 sec.; 4, M. Long (Lotus-Ford), 14 min. 55·8 sec.; 5. C. Crichton-Stewart (Cooper-B.M.C.); 6. K. J. St. John (Lotus-B.M.C.); 7, Dr. S. A. Goodwin (Lola-B.M.C.); 8, J. Berry (Lotus-Ford). **Fastest lap:** J. Y. Stewart and W. Banks, 54·8 m.p.h.

Grand Touring Cars up to 1,600 c.c. (10 laps)
Class A, 1,151-1,600 c.c.: 1, J. Y. Stewart (Lotus Elan), 9 min. 32·8 sec., 84·85 m.p.h.; 2, P. Arundell (Lotus Elan), 9 min. 35·0 sec.; 3, M. Beckwith (Lotus Elan), 9 min. 45·8 sec.; 4, M. H. Spence (Lotus Elan), 9 min. 53·0 sec. **Fastest lap:** J. Y. Stewart (Lotus Elan), 55·2 sec., 88·04 m.p.h. (new class record). **Class B, up to 1,150 c.c.:** 1, M. E. Garton (Austin-Healey Sprite), 10 min. 34·8 sec., 76·56 m.p.h.; 2, J. Harris (Austin-Healey Sebring Sprite), 10 min. 35·0 sec.; 3, T. D. Simpson (Marcos GT), 1 lap behind.

Guards Trophy (sports cars over 1,150 c.c., 20 laps)
Class A, over 2,000 c.c.: 1, J. Clark (Lotus-Ford), 18 min. 11·6 sec., 89·05 m.p.h.; 2, R. F. Pierpoint (Attila-Ford), 18 min. 32·2 sec. No other finishers. **Class B, 1,151-2,000 c.c.:** 1, R. Nathan (Brabham-Climax), 18 min. 43·4 sec., 86·52 m.p.h.; 2, J. Sutton (Lotus-Ford), 18 min. 46·2 sec.; 3, G. Breakell (Lotus-Ford), 19 laps; 4, W. J. Stein (Lotus-Ford), 19 laps. **Fastest lap:** J. Clark (Lotus-Ford), 52·8 sec., 92·04 m.p.h. (new class record).

Grovewood Trophy (Formula 2 cars, 30 laps)
1, J. Clark (Lotus), 26 min. 49·6 sec., 90·58 m.p.h.; 2, P. Arundell (Lotus), 26 min. 51·0 sec.; 3, J. Rindt (Brabham), 26 min. 57·8 sec.; 4, A. B. Rees (Brabham), 27 min. 00·6 sec.; 5. A. Hegbourne (Cooper); 6, D. Hulme (Brabham); 7, D. Hobbs (Merlyn); 8, A. Maggs (Lola). All engines Cosworth-Ford. **Fastest lap:** Clark, 52·6 sec., 92·39 m.p.h. (F2 class record).

Slip Molyslip Trophy (touring cars up to 1,300 c.c., 10 laps)
1, J. Fitzpatrick (Austin Mini-Cooper S), 10 min. 26·4 sec., 77·59 m.p.h.; 2, J. Handley (Morris Mini-Cooper S), 10 min. 29·6 sec.; 3, H. W. Ratcliffe (Morris Mini-Cooper S), 10 min. 42·8 sec.; 4, R. N. Cluley (Austin Mini-Cooper S), 10 min. 54·2 sec.; 5. B. Maskell (Austin Mini-Cooper S); 6. K. Costello (Austin Mini-Cooper S). **Fastest lap:** Fitzpatrick, 1 min. 00·8 sec., 79·93 m.p.h. (new class record).

With eight S-type Mini-Coopers and one near-standard Ford Anglia, Race 5 was a less exciting affair than Mini affrays generally are. John Fitzpatrick in a Downton-tuned S got out ahead, being pursued but never caught by J. Handley, R. W. Ratcliffe and R. N. Cluley; Fitzpatrick's pace was such that he set a new up to 1,300 c.c. class record at 79·93 m.p.h. The Ford retired.

The acceleration of the 4·7 litre Ford-engined Lotus 30 out of the hairpin drew gasps from the spectators.

Morris Mini-Cooper 1275 S 1,275 c.c.

THERE are many advocates of large car-engines who feel that capacity is the only way to get performance. They dislike little engines revving hard and noisily with sharply peaked power curves and little bottom-end torque. There is certainly a strong case for an engine which can slog along in a high gear if the mood takes one, without the fuss of keeping the engine speed up through constant gear changing.

When the latest two variations of engine size for the Mini Cooper S were announced last March, it was the 1,275 c.c. version which sounded the most interesting. This is the largest-engined Mini ever produced for sale and although the step was taken more with an eye on the up-to-1,300 c.c. class in international competition, the standard car is eminently suitable to ordinary road use nevertheless. Since the only significant difference between this car and the 1071 S we previously tested is the engine size, we have abbreviated our comments on the other aspects.

The story of the Mini-Cooper S engine, briefly, is very simple. Developed from the succesful formula Junior power unit, the components were put into quantity production to recuperate some, at least, of the high development costs. The crankshaft is made from high-tensile (EN 40B) steel, nitrided for surface hardness and extra strength, with 2·0in. dia. big ends and main bearing journals. All the S range of engines have a 70·6mm bore size, the variations in capa-

city being adjusted by the crank throw. The smallest engine (970 c.c.) has a bore-stroke ratio of 1·14 to 1, the middle one 1,071 c.c.) 1·03, whereas the largest one we have been testing is under-square with a ratio of 0·87 to 1.

Other extra-durable mechanical features common to all the engines include special pistons and con-rods, with substantially fatter gudgeon pins than the normal Cooper, Nimonic valves running in special copper-nickel guides operated by forged steel rockers, and a new camshaft giving a wide, flat torque curve.

All the engines are fitted with twin S.U. carburettors type HS2 (1·25in. dia.) and the 1275 S runs with a compression ration of 9·75 to 1. This largest-engined car has a higher final drive ratio than all the others—3·44 in place of 3·76 to 1—since the extra torque can overcome any losses in acceleration.

PRICES		£	s	d
Two-door saloon		625	0	0
Purchase tax		130	15	5
	Total (in U.K.)	755	15	5
Extras (Including P.T.)				
Wide-based wheels			3	0	5

How the Morris Mini-Cooper 1275 S compares:

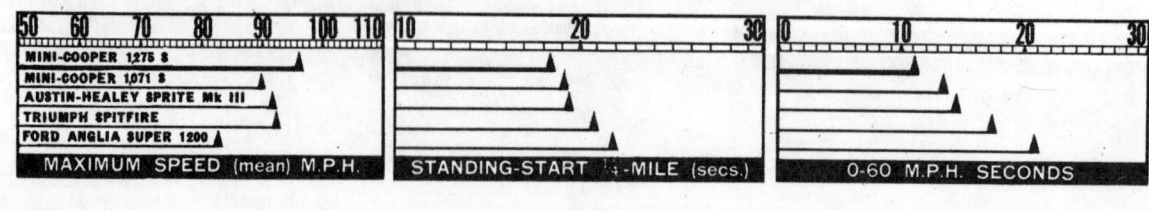

	MAXIMUM SPEED (mean) M.P.H.	STANDING-START ¼-MILE (secs.)	0-60 M.P.H. SECONDS
MINI-COOPER 1275 S			
MINI-COOPER 1071 S			
AUSTIN-HEALEY SPRITE Mk III			
TRIUMPH SPITFIRE			
FORD ANGLIA SUPER 1200			

Autocar road test • No. 1987

Make • MORRIS Type • MINI COOPER 1275 S (1,275 c.c.)
(Front engine, front-wheel drive)

Manufacturer : Morris Motors Ltd., Cowley, Oxford

Test Conditions

Weather.........................Showery, with no wind
Temperature......................16 deg. C.(60 deg. F.)
Barometer.....................................29·58in. Hg.
Dry concrete and tarmac surfaces.

Weight

Kerb weight (with oil, water and half-full fuel tank)
12·8 cwt (1,435lb-651kg)
Front-rear distribution, per cent F, 62·5; R, 37·5
Laden as tested ....15·8 cwt (1,771lb-806kg)

Turning Circles

Between kerbs............L, 31ft 11in.; R, 32ft 10in.
Between walls .... L, 32ft 8in.; R. 33ft 7in.
Turns of steering wheel lock to lock..............2·3

FUEL AND OIL CONSUMPTION

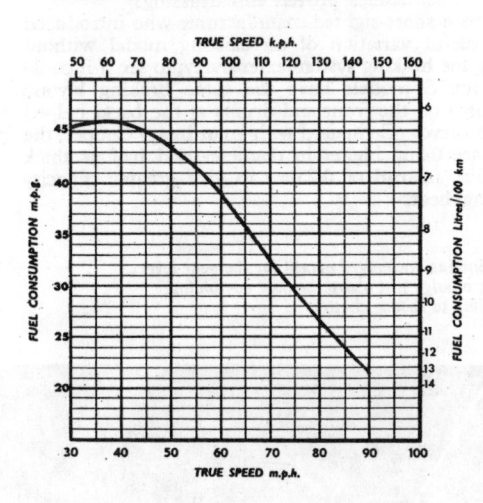

FUEL....................Super Premium Grade
(100–102 octane RM)

Test Distance .... 1,096 miles

Overall Consumption............28·5 m.p.g.
(9·9 litres/100 km.)

Estimated Consumption (DIN) 29·4 m.p.g.
(9·6 litres/100 km.)

Normal Range .... 27–35 m.p.g.
(8·1–10·5 litres/100 km.)

OIL: SAE 30.............Consumption 980 m.p.g.

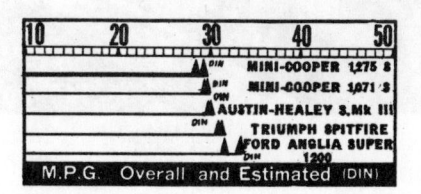

M.P.G. Overall and Estimated (DIN)

MAXIMUM SPEED AND ACCELERATION TIMES

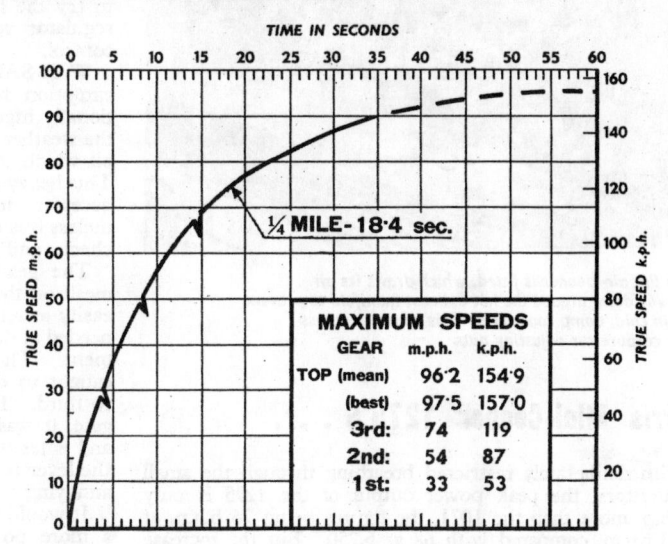

¼ MILE – 18·4 sec.

MAXIMUM SPEEDS		
GEAR	m.p.h.	k.p.h.
TOP (mean)	96·2	154·9
(best)	97·5	157·0
3rd:	74	119
2nd:	54	87
1st:	33	53

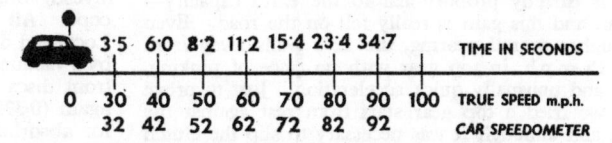

	3·5	6·0	8·2	11·2	15·4	23·4	34·7		TIME IN SECONDS
0	30	40	50	60	70	80	90	100	TRUE SPEED m.p.h.
	32	42	52	62	72	82	92		CAR SPEEDOMETER

Speed range, gear ratios and time in seconds

m.p.h.	Top (3·44)	Third (4·67)	Second (6·60)	First (11·02)
10—30	8·2	5·5	4·0	2·9
20—40	7·5	5·4	3·8	—
30—50	7·5	5·4	4·4	—
40—60	8·3	6·0	—	—
50—70	9·4	7·5	—	—
60—80	12·3	—	—	—
70—90	17·5	—	—	—

BRAKES	Pedal load	Retardation	Equiv. distance
(from 30 m.p.h. in neutral)	25lb	0·50g	60ft
	50lb	0·90g	33·5ft
Handbrake		0·30g	100ft

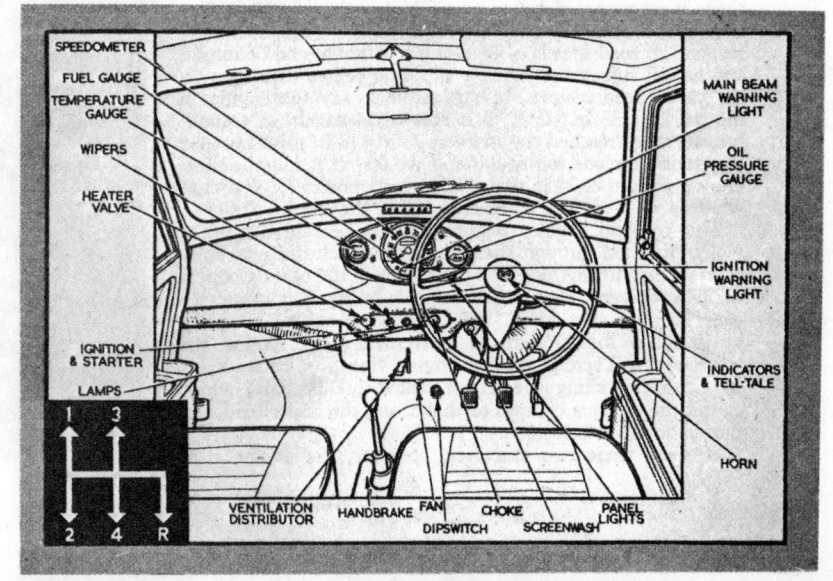

*A new flat air-cleaner is fitted, which draws its air
from way down around the hot exhaust manifold to prevent
icing in cold, damp conditions. This restricts access
to the carburettor adjusting nuts*

Morris Mini-Cooper 1275 S . . .

With deliberately restricted breathing through the small carburettors, the peak power output of the 1275 is only 8 b.h.p. more than the 1071, the figures being 76 b.h.p. at 5,900 r.p.m. compared with 68 at 5,750. But the increase in torque is directly proportional to the extra capacity—27 per cent, and this gain is really felt on the road. Even with the higher overall gearing, the 1275 pulls away from as low as 15 m.p.h. in top gear with no trace of pinking, no judder, and unusually quick acceleration. Just to prove the point, we tried a top gear start from rest against the stopwatch; and although it was necessary to slip the clutch until just over 10 m.p.h., thereafter the car pulled strongly to record a time of 23·0sec from 0 to 60 m.p.h. and 25·4sec for the standing start quarter-mile.

Taken off the line in a more spirited fashion, using all the gears, the corresponding times from rest to 60 m.p.h. and for the quarter-mile were only 11·2sec and 18·4sec. (Figures for the 1071 we tested on 12 April 1963 were 13·5sec and 19·2sec respectively.) With a car of this nature, the exact change points for maximum acceleration are not always obvious, and as the speedometer is not very well positioned for the driver to see, we installed a Smiths electronic rev. counter alongside it.

Using this instrument we tried a number of varying techniques, making changes above the power peak at 6,300 r.p.m., on the peak at 5,900 r.p.m. and below it at 5,700 and 5,500 r.p.m.. Because of the gearbox spacing, the best times were recorded by going no higher than 5,900 in first and second, and only 5,700 r.p.m. in third. These revs correspond to road speeds of 30 m.p.h., 50 m.p.h. and 66 m.p.h. and are all substantially below the little yellow markings on the car's speedometer. It is possible to rev the engine to the mark at 33 m.p.h. in first and to 54 m.p.h. in second, but we never reached the mark at 78 m.p.h. in third because the engine became too rough and we feared it might suffer. Our maximum of 74 in this gear corresponds to 6,200 r.p.m., whereas we were able to reach 6,500 in the lower gears.

This top-end engine roughness could be felt as quite a violet vibration through the gear lever, which did not, however, chatter to anything like the extent of the other Coopers we have driven. One reason for this is that a new pattern gear lever is now fitted, with a bonded rubber coupling at its base to dampen its movement and reduce the resonances. There is another engine vibration, corresponding with the idling speed just below 1,000 r.p.m., which causes the steering column to shake, and the wiper blades to flutter on the windscreen.

Between these two extremes, however, the engine runs with all the sweetness of the shorter-stroke units, and the extra crankshaft flexibility is not noticed. There are no flat-spots—even when cold, the car always firing first turn of the starter with no choke, and pulling strongly at once.

During sustained high-speed running on motorways the oil pressure dropped from its customary 60 p.s.i. to about 45 p.s.i., indicating the need for an oil cooler under these conditions. The water temperature also approached the danger zone unless the speed was kept below 85 m.p.h. Although the weather throughout the test was too warm to try the heater, the temperature control now has a new regulator valve which appears to give more progressive control.

With SAE 10W/30 multigrade oil in the sump, the consumption during the first 500 miles of the test was exceptionally high—725 m.p.g. or only 90 miles per pint. Since the weather was warm, we then changed to a straight SAE 30 oil which improved the consumption slightly to 980 m.p.g. This heavy rate can be accounted for by the larger clearances necessary in an engine of high specific output, but nevertheless it is extremely inconvenient to make repeated oil level checks and add some at each fuel stop—every 140 miles.

The gearbox works very well, with powerful synchromesh on the upper three ratios. First gear always went in easily at rest, and even on the move no particular skill was needed at double-declutching for a satisfyingly silent engagement. There was just one little fault which sometimes caught us out when hurrying the change up from second to third. If the lever was moved too forcefully across the gate, it was possible to over-ride the reverse spring catch and enter the blank "slot" ahead of reverse itself. Bringing the lever back to neutral restored things very easily, but this annoying trick sometimes proved embarrassing.

It would be a short-sighted manufacturer who introduced a more powerful variation of an existing model without investigating the braking system extensively, to see if it could cope. All the S models have the same braking layout, Lockheed discs on the front and drums at the back, helped by a vacuum servo. Compared with an ordinary Cooper, the front discs are 0·5in. bigger in diameter and half as thick again (0·375in. instead of 0·25in.) to give greater capacity for absorbing heat,

*The interior remains unchanged except for the gear lever,
which now has a bulge just above the floor containing
a rubber coupling to reduce chatter*

As well as carrying out our usual test of retardation, in relation to pedal pressures from 30 m.p.h. in neutral, we did a brief fade test. After six stops from over 75 m.p.h. in quick succession the front discs were smoking heavily, but still recording 0·75g with only 50lb load on the pedal. From 30 m.p.h. with the brakes cold, it took only a little over 50lb to lock all four wheels together, our best retardation with the wheels not quite locked being 0·9g.

Much has already been written in various tests on the ride comfort and characteristics of the whole Mini family. The interesting thing about this latest Cooper S is the way, in the dry at least, its cornering powers can be abused almost without danger. Several times on a closed test track, we deliberately lifted off the accelerator at the apex of a sharp corner at 65 m.p.h. or so, and the only change in the little car's attitude was from mild understeer to a neutral characteristic. Snapping open the throttle again caused the front of the car to pull outwards from the chosen line, but this tendency was easily overcome by applying more lock. The car was shod with Dunlop SP41 tyres (C41 tread pattern extending right round the shoulders of an SP rigid-breaker carcass) on the optional wide-rimmed wheels and, in this aspect of stability under varying power conditions, during cornering it was by far the best Mini we have ever driven. Rapid getaways and wheelspin in corners caused a deep-noted howl, but in ordinary running the tyres were outstandingly quiet and free from road roar.

In almost any degree of traffic, country roads, trunk routes or city streets, the 1275 S is one of the quickest ways of getting from A to B in safety. Far from a quart squeezed into the proverbial pint pot, this car has handling and braking well within its high standards of performance.

All Cooper models have two-colour p.v.c. upholstery and carpets on the floor. The seats are not shaped to give support while cornering, and their backrests have little rake

Specification : Morris Mini-Cooper 1275 S

ENGINE

Cylinders ...	... 4, in-line
Bore ...	... 70·64mm (2·78in.)
Stroke ...	... 81·33mm (3·20in.)
Displacement ...	... 1,275 c.c. (77·9 cu. in.)
Valve gear ...	... Overhead, pushrods and rockers
Compression ratio	9·75-to-1
Carburettors ...	... S.U. Twin HS2
Fuel pump ...	... S.U. electric AUF
Oil filter ...	... Hoburn Eaton, full flow, renewable element
Max. power ...	... 76 b.h.p. (net) at 5,900 r.p.m.
Max. torque ...	... 79 lb. ft. at 3,000 r.p.m.

TRANSMISSION

Clutch ...	... Borg and Beck diaphragm spring 7·125in. dia.
Gearbox ...	... Four speed, synchromesh on 2nd, 3rd and top
Gear ratios ...	... Top 1·00, Third 1·36, Second 1·92, First and Reverse 3·20
Final drive ...	... Helical spur gears, 3·44 to 1

CHASSIS

Construction	... Integral with steel body

SUSPENSION

Front ...	... Independent, wishbones, rubber cone springs, telescopic dampers
Rear ...	... Independent, trailing arms, rubber cone springs, telescopic dampers
Steering ...	... Rack and pinion Wheel dia., 15·75in.

BRAKES

Type ...	... Lockheed, disc front, drum rear, vacuum servo
Dimensions ...	... F. 7·5in. dia., R. 7in. dia., 1·25in. wide shoes
Swept area ...	... F. 120 sq. in.; R. 55 sq. in. Total: 175 sq. in (222 sq. in per ton laden)

WHEELS

Type ...	... Ventilated steel disc, 4 studs, 4·5in. wide rim.
Tyres ...	... 145—10in. Dunlop SP41

EQUIPMENT

Battery ...	... Lucas 12-volt 43-amp. hr.
Headlamps ...	... Lucas 50-40 watt
Reversing lamp ...	... Extra
Electric fuses ...	... 2
Screen wipers ...	... Single-speed, non-parking
Screen washer ...	... Standard, manual plunger
Interior heater ...	... Standard, fresh-air type
Safety belts ...	... Extra, anchorages provided
Interior trim ...	... P.V.C., plastic headlining
Floor covering ...	... Carpet, felt underlay
Starting handle ...	... No provision
Jack ...	... Screw pillar
Jacking points ...	... One each side under door sill
Other bodies ...	... None

PERFORMANCE DATA

Top gear m.p.h. per 1,000 r.p.m.	16·1
Mean piston speed at max. power	3,140ft/min.
Engine revs. at mean max. speed	6,040 r.p.m.
B.h.p. per ton laden	95

MAINTENANCE

Fuel tank ...	... 5·5 Imp. gallons (no reserve
Cooling system ...	... 6·25 pints (inc. heater)
Engine sump and transmission ...	... 8·5 pints. Change oil every 6,000 miles; change filter element every 6,000 miles
Grease ...	... 8 points every 3,000 miles
Tyre pressures ...	... F. 28; R. 26 p.s.i. (normal driving)

▼ *Scale: 0·3in. to 1ft. Cushions uncompressed.*

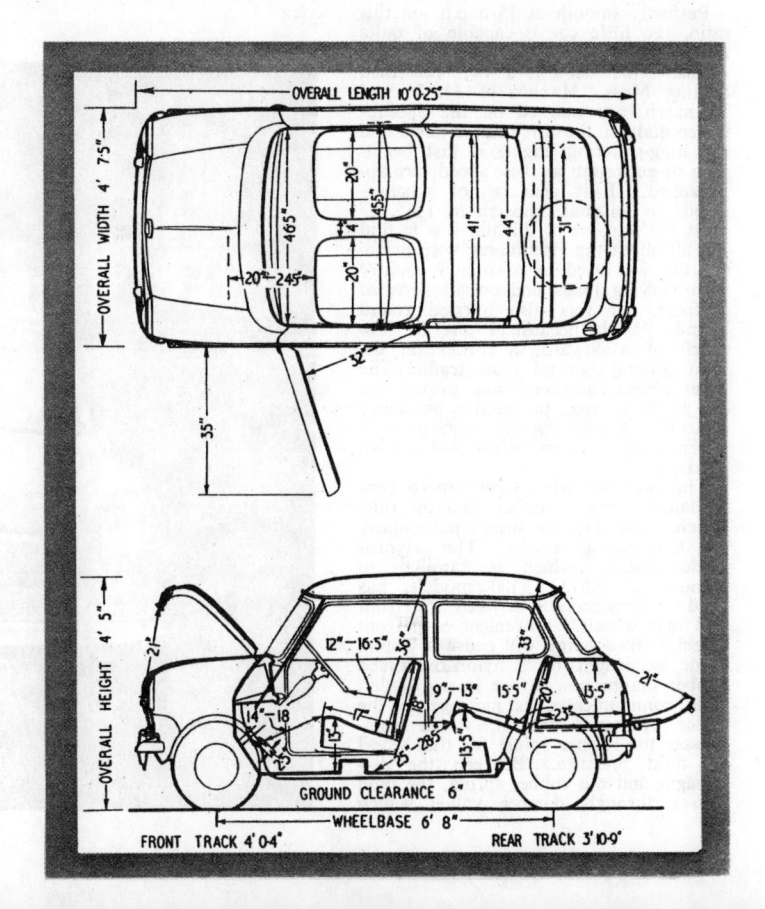

OVERALL LENGTH 10' 0·25"
OVERALL WIDTH 4' 7·5"
OVERALL HEIGHT 4' 5"
GROUND CLEARANCE 6"
WHEELBASE 6' 8"
FRONT TRACK 4' 0·4"
REAR TRACK 3' 10·9"

A<small>UTOSPORT</small> has already conducted a full road test of the S 1275; indeed, it was one of our outstanding experiences of 1964. Now, the Hydrolastic version has come along, which has given us an excuse to enjoy this delicious little machine for another week.

A full and accurate list of performance figures, together with a graph, were given with the A<small>UTOSPORT</small> road test (10th July, 1964). In this instance, no attempt has been made to better these figures, as the weather was against such a project. However, it was established that the latest model, with Hydrolastic suspension, has about the same performance as its predecessor, which one would expect. In brief, the car is

MINI-COOPER
S 1275
with
HYDROLASTIC SUSPENSION
Road impressions by JOHN BOLSTER

ridiculously quick up to 70 m.p.h. or so when the unstreamlined form of its family saloon body begins to be apparent. Nevertheless, a timed 100 m.p.h. is available and as much as 104 m.p.h. may be touched briefly. The Cooper seems happy at a 90 m.p.h. cruising speed.

The acceleration is quite remarkable, a time of 9 secs. for the standstill to 60 m.p.h. range being oustanding. What is perhaps even more spectacular is the top-gear flexibility.

Perfectly smooth at 15 m.p.h. on this ratio, the little car is capable of quite fierce acceleration from 25 m.p.h. onwards. "It's just like a V8," remarked Stirling Moss. Maxima of 34, 55 and 78 m.p.h. are marked on the speedometer dial for the three lower gears and the long-suffering engine is just beginning to get rough as these speeds are approached. First gear is not synchronized, an unusual omission in 1965.

As the purpose of this test was to find out all about the Hydrolastic suspension, the car was used in town and country with varying loads and on all sorts of surfaces. It was also hurried round Brands Hatch because I am far too much of a coward to corner on the limit among normal road traffic. The main object, however, was to use the car as the average prospective purchaser and his family would, regarding it alternatively as a sports car and a town carriage.

The ordinary Mini gave superb controllability but a rather choppy ride, which could become tiring, particularly on Continental roads. The Hydrolastic system, which is familiar to owners of B.M.C. 1100 models, has fluid interconnection between the front and rear wheels. Movement of a front wheel over a bump will cause a piston, made watertight by a nylon-reinforced rubber diaphragm, to rise in a cylinder. This communicates its message to the water contained therein, which compresses the rubber spring for that wheel by fluid pressure. Between the diaphragm and the rubber spring, the fluid passes through damper valves which

take the place of separate shock absorbers.

In addition to acting as a link between the front wheel and its own spring, the fluid is also conveyed by a pipe to the rear suspension unit on the same side, which is similar to the front assembly. Thus, if the front wheel rises over a bump it causes the rear of the car to rise a little, and vice versa. This greatly reduces the movement known as pitching, which is the most objectionable motion that a car can have. Thus, the Hydrolastic Mini is not outstandingly softly sprung but it is almost immune from the pitching movement which afflicts small cars particularly, therefore, it combines stability with comfort to an extent not previously approached by orthodox designs. Additional small helical springs at the rear have a steadying effect.

On the road, the Hydrolastic Mini feels almost soggy at parking speeds but the suspension seems to become progressively harder as the velocity increases. The improvement in riding comfort is very great, the Hydrolastic car giving a much more level motion, both when the driver is alone and when all the seats are occupied. At racing speeds on a circuit, the new car is just as marvellously controllable as last year's model was. Indeed, I felt even more at home in it. No doubt we shall soon see, in saloon car racing, whether the Hydrolastic system can be made to give increased cornering power or not. For use on the road, though, I am completely sold on it, from every point of view.

In spite of the excellence of the suspension, I would still like more comfortable seats, and the steering column is rather too upright for me. These are things that an enthusiastic owner can have altered to his own ideas, but we are speaking of the car in standard form. The fuel tank may also be duplicated and I would certainly specify this. Yet the car is by no means heavy on petrol. If a speed of 60 m.p.h. is not exceeded, the S 1275 will return better consumption figures than a standard Mini.

At the other end of the scale, lots of fast work on the motorway will bring the consumption up to 27 m.p.g. or so. During my whole test, which included town work, shopping, fairly fast driving in the country, some flat-out dicing at Brands Hatch, but no long journeys on the motorway, I averaged just over 30 m.p.g. I used very little oil, but in hot weather on the M1 this might be a different story. Incidentally, no car could start more quickly on a cold morning.

In practice, the Mini-Cooper is even faster than its performance figures would imply. This is because it gives tons of power at all speeds, and simply jumps away on any gear at a touch of the pedal. I do hope that it will not gain too much of a reputation as a "racer", for it is an even better car for auntie than her standard model. One is not a boy racer just because one enjoys the extra safety of rapid overtaking, and fade-free braking, and the sure roadholding of SP tyres on wide rims—insurance companies please note.

There will always be a demand for expensive, hand-made cars, and those who can afford something really exclusive are delighted to pay an unecomonic price. Yet, the Mini-Cooper S 1275 presents safe, high-speed motoring on the road which few of these costly toys can equal. At £778, including purchase tax, it is a product of which Britain can be proud.

FLYING FINN LEADS BMC HOME

Timo Makinen of Finland

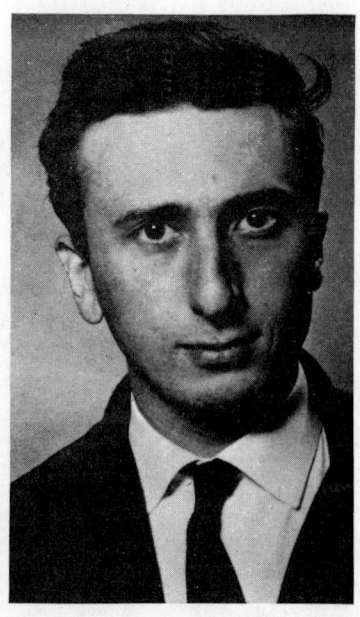

Paul Easter, a Buckinghamshire engineer

With a second consecutive win in the Monte Carlo Rally, B.M.C. confirmed their position as the world's top rally team, having now won virtually every major rally in the calendar

FOR the second time in two years B.M.C. Minis have snatched the most glamorous prize in motor sport—an outright win in the Monte Carlo Rally. Following in the footsteps of Paddy Hopkirk and Henry Liddon in 1964, Timo Makinen and Paul Easter in a 1275 c.c. Mini-Cooper 'S' established a clear lead on the 2,600-mile run to Monte Carlo and then retained it over the 400-mile special Mountain Circuit on the final night.

Only 35 of the 276 starters reached Monte Carlo from nine starting points and of these three were official B.M.C. team entries and three others, a Cooper 'S' and two 1800s, were private entries. The Mountain Circuit took a further heavy toll and another 11 cars failed to complete the night run. This, however, only decided the final placings and failure to finish this special stage did not prevent cars winning their classes.

According to competitors this was one of the most difficult Monte's in post-war years. Bad road conditions and heavy snow made driving hazardous and only one car, that of Makinen and Easter, arrived at Monte Carlo with a clean sheet. All the other 34 finishers had picked up penalty points.

When the rally proper ended at Monte Carlo Makinen had a 10-minute lead over Bianchi who was lying in second place in a Citroen. As well as having a clean sheet Makinen and Easter also recorded the fastest times on many of the special stages into Monte Carlo. Because the organizers had doubled the penalty system for the Mountain Circuit Bianchi had, in fact, to catch up 20 minutes on Makinen to win. Bianchi was being hotly pursued by Bohringer, driving a Porsche, who was only two marks down on him. As it turned out neither Bianchi or Bohringer were able to make much impression on Makinen on the Mountain Circuit and he finished a clear winner of the Rally and, of course, of his class. Bianchi crashed in the mountains in his efforts to catch Makinen and Bohringer took second place.

In addition to having the outright winner B.M.C. also gained the first three places in Category 1, Class IV, with three other Coopers. These were driven by Hopkirk/Liddon, the Morley twins, and Latrobe/Bailley. For his performance in being the highest placed British private entry John Latrobe is almost certain to win the *Autosport* Trophy.

Although B.M.C. had six cars in at the finish, because a car from each of the official teams had dropped out the Manufacturers Team Prize eluded them. This went to Swedish Saabs with Citroens second. The anticipated strong challenge from American Ford, with their team of Mustangs, did not materialize. On the other hand one of the surprises of the Rally was the performance of the two 1800s, only announced by B.M.C. six months ago and completely untried in rallies. This is even more noteworthy as both these cars were private entries.

Above. Timo Makinen and Paul Easter on their way to an outright win in the Monte Carlo Rally

Below. In spite of a broken tie-bar, which had to be welded three times, Paddy Hopkirk and Henry Liddon still won their class

Top Donald and Earle Morley, also in a Mini-Cooper 'S', took second place to Hopkirk/Liddon in their class

Left. John Latrobe, driving a privately entered Mini-Cooper 'S', gained third place in the same class to give B.M.C. 1—2—3. Latrobe may also win the *Autosport* Trophy awarded to the best-placed British private entry

Below. Raymond Joss/Fitzpatrick in one of the two privately entered 1800s. Both finished the course taking 29th and 31st places overall

Photo Michael Cooper

MINIS IN COMPETITION

THE 1959 RACING SEASON was more than half-completed when B.M.C. announced the Mini. In those days, which seem a long way off now, small-car saloon racing was dominated by other B.M.C. products—and in particular the fabulous A40 of Doc Shepherd. Come to that, he made his mark in the big-car world as well—remember those fantastic races at Brands Hatch, when Shepherd used to be up amongst the Jaguars? In the world of international rallying, Saab was beginning to dislodge Renault.

Before the season was out, that self-same Doc Shepherd had got his hands on a Mini and, at the tail end of the year, won his class with it in a Snetterton M.R.C. sprint, possibly the first actual competition success of the car. If it was, he started something. A private team of enthusiasts under the name Cambridge Racing, which in those days was a pretty successful organisation, added a Mini to their team strength for the 1960 season, running it as third string to a brace of A35s. The Bournemouth Rally, run in November by the West Hants. and Dorset Car Club, saw a team of Minis winning the team prize and for the R.A.C. Rally, at the end of the month, the eight-car B.M.C. assault included three of them, driven by Ken James, Alex Pitts and Pat Ozanne: there were three private entries, too, but the Minor 1000 seemed to be more popular—Pat Moss had one that year. Highest Mini in the results of the car's first international was thirteenth, with a second place in their class. In December, B.M.C. put two of them in for the Portuguese Rally while, in a minor key (perhaps a Mini-Minor key?) an Austin Seven, which was the name given to the Austin version of the two identical models (they used, if you remember, to spell it Se7en in the most unpronounceable way) won the premier award in an Irish trial. In the Tour of Corsica another

Mini took eleventh place, but the event was still won by a Dauphine.

The Boxing Day Brands Hatch meeting ended the year so far as motor sport was concerned. Several of the Minis turned out for the small saloon car race, but only one finished: John Whitmore's went exceedingly fast until a puncture put him out, but, says a contemporary report, the others "were less impressive". At the Racing Car Show—the first of all, in January 1960, several of the leading tuning firms exhibited equipment for the ADO15, as in-people were beginning to call it, but as yet there was no sign of the swamping of the field which became evident later on. There were half-a-dozen, though, in the Monte Carlo Rally of 1960, including several among the works team, which suggested that the cars' potential was beginning to be realised. But it was early days: one of the works Minis in that event was driven by the Morley brothers, who were described in a contemporary as "up-and-coming", as indeed they certainly were. The same foresight was not, however, extended to their mount, and reports of the day tended to concentrate on the Ford Anglia 105E, announced at about the same time and in those days more popular with the tuners. As it turned out the Morleys finished well up among the Mini crews, in 33rd place. Of the six entries, all finished, four got into the final classification test and the highest-placed car was that of Peter Riley and Rupert Jones, in 23rd place. It was beginning to happen.

As the 1960 season got steadily under way, Minis did well in the Tulip Rally and produced some startling results in the touring car race at the May International meeting at Silverstone. The British Racing and Sports Car Club was running a special championship for "SupaTura" cars this year and by June, with half the season gone, no-one was surprised to find Doc Shepherd's A40 in the lead. But in seventh place was the first of the Minis . . .

By the end of the year Minis were growing in popularity in all directions. The second Racing Car Show, however, in January, 1961, was still a long way from being dominated by Mini parts. The Monte Carlo Rally, which that year favoured small cars, saw Panhards taking the first three places, and the best Mini was lower down the list than in the previous year's event. In club racing the Mini went faster and faster, although as yet it was not exactly beating all-comers, and the SupaTura championship went to Shepherd's A40.

The 1961 season, however, saw a change in all this. By the time the season opened at Snetterton with the B.R.S.C.C. spring meeting, there were enough Minis to run a special 850 class, and Mick Clare won from John Aley at 72·33 m.p.h. To put it another way, his race average for eight laps was about the maximum speed of the standard car—and these were 850s, don't forget. At Oulton Park the same weekend, Smith's Austin Mini won a closed car handicap—so far as we can trace one of the first actual race wins for the Mini. Back at Snetterton, the first British international of the year, the Lombank Trophy meeting, saw Hurrell's Saab win the Mini class from Doc Shepherd's A40: at Goodwood that Easter John Whitmore scored what may be the model's first victory at an international meeting by winning—indeed, running away with—the 1,000 c.c. class in the saloon car race, on a soaking wet track, at over 68 m.p.h.

In April, 1961, the B.R.S.C.C. showed an appreciation of the way things were going by holding the first-ever race for Minis only. This was an event of wild excitement, in the best Mini-racing traditions, and although won by Graham Burrows, at 62·5 m.p.h., from Doc E. H. M. Paul, the fight for the places went on throughout the race: this was one of the early occasions on which Christabel Carlisle drew the admiration of the crowd with her spirited driving. At Aintree, the "200" meeting included a saloon car race, and John Whitmore won the Mini class from John Aley at 66 m.p.h.; after a tremendous dice between Aley and Doc Shepherd the Doctor's Mini skated around on its roof for some distance. No-one was hurt, and he had the consolation of fastest lap at 68 m.p.h.—and Mike Parkes' fastest lap of the race, in a 3·8 Jaguar was only 77 m.p.h.! And so it went on. More and more Minis, going faster and faster. But even the factory hadn't finished yet. In September, 1961, they announced the Mini-Cooper, with a 997 c.c. engine and 55 b.h.p. as standard. This set the cat among the pigeons, and had as much to do with the demise of things like the A40 as race-winning cars as anything. But their day was done, had they realised it, in 1959.

One of the best races of the year—in any category—took place at Snetterton, however, before the Cooper emerged. This was the Scott-Brown Trophy meeting, when Christabel Carlisle laughed at the weaker sex idea—a not uncommon habit of hers before she retired—and made the established Mini experts—Clare Aley and the rest—a very hard time keeping up. These three in particular were seldom in the same order for more than a few hundred yards at a time—racing right on the limit, weaving, in line astern, line abreast or in echelon. Not more than half-a-Mini-length covered the three as they crossed the line—fractionally Clare first, at 71·4 m.p.h., then Christabel and then Aley. There was another similarly closely-fought duel at Brands Hatch on August Bank Holiday, when John Whitmore led until he retired with a lack of gears; John Aley then took over until overheating forced him out and, eventually Hamlin led Clare across the line at only a decimal short of 70 m.p.h.: Hamlin made fastest lap, too, at 71·1 m.p.h. This sort of thing was going on at circuits up and down the country, and indeed all over Europe, just about every weekend. It was the hey-day of the 850 Mini. Occasionally, on the

A gaggle of Minis at Goodwood with the opposition on the grass

John Aley and Christabel Carlisle playing silly whatzits at Snetterton

continent, interlopers with B.M.W.s tried to get in on the act, but these, by and large, were swiftly dealt with.

The arrival of the Cooper, in September, was too late to make any difference, just as, in earlier days, the late-August arrival of the Mini-brick (as it was then affectionately known: the term seems to have died out nowadays) was too late to affect anything in 1959. At Snetterton, John Whitmore won the Mini race in his Don Moore-tuned car, harried all the way by Miss Carlisle, and won the B.R.S.C.C. Touring Car Championship.

By 1962, everybody who was anybody was wearing a 1,000 c.c. engine on his Mini. The 850s were still going strong, but the limelight tended to go more strongly onto the larger-engined cars. The 850 had been outstandingly successful—and it had all happened a bit suddenly, after a slow start. But the advent of the Cooper version left no-one unprepared—it simply meant that here was an engine with much of the basic work of tuning already carried out. For the next two years the Cooper-Mini carried the flag from success to success: the tuners got busy, and things began to happen.

It is obvious that a full list of competition successes of this car would fill several books the size of the London telephone directory. CARS AND CAR CONVERSIONS isn't as big as that and, anyway, we've only got a few pages. Let it rest that 1962 was a year of intensive development: 1963 was a year of tremendous international success. The ball started rolling with the Monte Carlo Rally, in January: Cooper-Minis finished third and sixth in general classification, took the first three places in their class in the touring category, and won their class in Grand Touring. In April, in the Tulip Rally, they took four of the first five places in the 1,000 c.c. class in the Touring category, including second overall in that category, were third overall in the G.T. category and won the 1,300 c.c. class.

Success was not by any means confined to rallies. At Spa in May, the 1-litre class of the international touring car race saw Cooper-Minis in the first three places and in July, they took the first three places in their class, and the team prize, in the

"Motor" international Six Hours Race at Brands Hatch. They won their classes in the touring and G.T. categories of the Midnight Sun Rally, they won the Coupe des Dames and a Coupe des Alpes, not to mention the Manufacturers' Team Prize. In the R.A.C. International Rally, in November, Cooper-Minis took a one-two in the 1-litre class, and third in the 850 c.c. class. And so on, and so on.

Meanwhile, in April, 1963, B.M.C. had introduced the first of the Cooper "S" models, the now-discontinued 1,071 c.c. version. These had, of course, much more power, even in standard form, than the ordinary Cooper, were fitted with improved disc brakes, close-ratio gears and so on—the full works, as you might say. Right from the word go the move paid off handsomely: in June, Aaltonen/Ambrose won outright the Touring Category of the Alpine Rally, which is a fair old event to finish, much less win an award. At the model's first racing appearance, at the British Grand Prix meeting at Silverstone the following month, it took the first five places in the 1,300 c.c. class of the Touring Car Race, driven by John Whitmore, Paddy Hopkirk, Christabel Carlisle, Mick Clare and John Fenning, in that order. Abroad, it won the 1,300 c.c. class in a race at Roskilde Ring, in Denmark and with Hopkirk and Liddon, in September, a Cooper "S" cleaned up in the Tour de France by winning the touring category index, third overall in the touring category (scratch) and first in its class.

The story goes on like that, and repetition—even if its repetition of victory and success—grows tedious. We'll just say that the pace of winning grew hotter and hotter until in January, 1964, it achieved something which few other cars have achieved in so short a time—an outright win in the Monte Carlo Rally. This, actually, was only half the story. Paddy Hopkirk's brilliant win (not forgetting Henry Liddon, his co-driver) was accompanied by a total of three of the first seven places overall; the first three places in the class, and the Manufacturers' Team Prize. Not bad for a car which had been in existence for less than five years, and possibly a unique record of success in one of the greatest of the international rallies.

A fabulous Mini battle; John Fenning leading Sir John Whitmore: Goodwood

THE SOOPER COOPER

BMC has come up with the locally-made version of the Mini-Cooper 1275S—and does it go!

MOST people have known for some time that BMC was building a local version of the very fast Mini-Cooper 1275S-type, but few expected it to be as dynamic as it is. The car, still not generally released, has been designed all over again for Australian usage, and is — even in its present unproven stage — a much better car than the British version.

Originally, BMC was to market the car for £1140 including tax, a staggeringly-low figure considering that the British car with equivalent specifications costs £1650 fully imported. But the Sydney manufacturer was able to rationalise a lot of its costing by basing the car on the Mini De Luxe panels and running gear. Where the rest of it went is a bit of a mystery.

The car we drove, two months before release (that's it jumping joyfully on the cover), was an engineering prototype that conformed exactly to the version that is to be sold, except that the Hydrolastic suspension was carrying slightly stiffer settings. Also, BMC at that time was considering a change from 1¼ in. to 1½ in. SU carburettors.

At the price, the Cooper S is possibly the cheapest genuine high-performance car you can buy. It is almost impossible to describe the sensations produced by this little box-shaped device; the only way to appreciate it is to drive it on the open road, and if any BMC salesman tries to sell one of these by

Inside, good carpet and good quality upholstery make things very comfortable. Speedometer carries rpm markings. Windup windows of course.

Brake servo at left takes up some space in engine bay. This was the prototype, with extra plumbing.

driving the client around a city block then he should be despatched immediately to the Home For Old And Infirm Salesmen, if there is one. Not that the car does not perform in traffic; in fact, the only faster way through city and urban traffic than this is by helicopter. But its real character does not show up until the derestriction signs are posted.

During our two days with the car we several times saw 110 mph indicated on slight downhill stretches. A future full road test — with proper speedometer corrections — will verify this or make idiots out of us, but we feel certain that this is a true 100 mph car. Now, 100 mph in a Mini-bin is not very different from 60 mph in a Mini-bin, but if you consider that this is a road car, selling for less than £1500, and is essentially a derivation from an 848 cc family potterer than never saw more than 73 mph in its life, then the truth comes home.

The Australian Cooper has the same body as the De Luxe, including the windup windows and quarter-vents unique to this country, and Hydrolastic suspension. The suspension is on harder settings than used for the De Luxe, but the ride is only slightly sharper. From outside, it looks little different to the 998 cc Mini-Cooper, except that the smaller-engined version still has sliding windows.

But look closely; you discover wider rims all round, a slightly lower and more aggressive "sit", and the small chromed "S" badge on the boot lid. Under the bonnet, however, is the husky 1275 engine flanked by a big brake servo and snaking with feeder pipes into an oil cooler. Inside, you find neat woven and bound carpet with a small rubber insert for the driver's heels, the good-quality pvc used in the now-defunct Wolseley 24/80, and a generally high standard of finish and trim. The wide parcels shelf supports a central nacelle housing speedometer/fuel gauge, and oil pressure and water temperature gauges. Beneath this is the sub-panel for the switches and beneath this again the rather ugly black-crackle-finish heater control box.

Otherwise the interior is much the same as the De Luxe, right down to narrower bins on the doors, self-parking wipers, short remote-control shift, seat rake angle adjustment, and even that ridiculously small and sharp-edged rear vision mirror.

But the buyers of Cooper S-types will be less concerned with the interior than the performance. The little thing accelerates like an artillery shell — so much so that we feel that a tachometer is an unfortunate (if economically justified) omission. The speedometer does have maximum rpm points in gears marked on its face, and these are accurate guides, so there you are. The full road test will show up just how fast the car does accelerate, but what figures will not demonstrate is just how much of this power is useable. Too many cars accelerate quite hard but have such low levels of adhesion or braking that they can use their available power perhaps only 50 percent of the time.

In the Cooper one needs such small openings and short stretches of road that every last cubic centimetre can be used fully. As a point-to-point car we can think of few that will stay with it, because it can make 90 mph overtaking manoeuvres in spaces that would horrify a good driver in a good Cortina GT, for instance. The handling of any Mini — as most of us know by now — has always given a driver a feeling of great confidence in his control of the car and the car's directional stability. This is exaggerated five times in the Cooper, so that you find yourself leaving narrower and narrower margins everywhere.

We can foresee a great demand for these little cars, and possibly an equally big business in swiping "S" badges to be used on Mini De Luxes. Never mind. They'll find the moment of truth when the derestriction signs pop up.   #

A dead ringer for the Mini De Luxe, until you pick those wider rims and the small "S" bezel at the rear.

Spare wheel sits under floor shelf in boot. Extra long-range tank is in the right wing.

These Seats Travel At 113 mph

A BROADSPEED £100 CONVERSION OF A COOPER 'S'

NOWADAYS the tuning industry is at the point where, apart from out-and-out racing, mere speed just isn't enough. The importance of reliability goes without saying and of course applies to racing too, but even for road use a properly converted car must show an improvement in all respects, and not just in performance, over the standard product.

All of which simply goes to convince us that the 1275 Cooper "S" we've just been testing is indeed a properly converted car. Look at the improvements. The chief weaknesses of the Cooper "S" are noise, vibration and, driven hard, an excessive oil consumption. The car we've been testing is quieter, suffers from practically no vibration despite its ability to survive a 2,000 r.p.m. increase in engine speed, and did about 200 miles per pint of oil—driven hard—compared with the 75 m.p.p. we've had from standard cars. Reliability? Well, the owner, Mr. P. Perrey, a Birmingham solicitor, has done about 19,000 miles in the car and is still sufficiently pleased with it to let us part him from it for a few days for the test, which argues that it's alright on that score. Improvements in performance? But yes. This car did 113 m.p.h. in one direction, and will get to sixty, two-up, in under nine seconds, which isn't hanging about. Given decent conditions it ought to be possible to improve on this—we were getting wheelspin in second gear as well as first, and it took us three days before we could find dry enough weather to make it worth trying.

And now for the 64-dollar question—who did the conversion? The answer, friends, is Broadspeed—to be exact, Broadspeed Engineering Ltd., of 101 Stratford Road, Birmingham 11. This car represents their Stage III tune—what is called their "Hundred Pound Conversion" on the 1275 Cooper "S". This is the one they suggest is suitable for the owner who wants to combine everyday motoring with a chance of success in week-end competitions. What they do is to modify the cylinder head, enlarging the inlet valve ports and tracts, polish ports and recontour the combustion chambers to their racing specification. The head is then machined and surface-ground to give a compression ratio of 10·5 to 1, and on to a modified manifold are mounted two modified 1½ in. S.U. H4 carbs, along with a special exhaust manifold. The camshaft is changed for the Broadspeed "Road/Race" pattern, and bob's yer uncle. Not included in the written specification, but there nevertheless, is the reason why the oil pressure is reduced drastically—special oil seals are fitted to the valve guides, and this it seems does the trick.

The test car had a number of other fixtures and fittings, and although these don't materially affect the performance they are worth mentioning. It had Cooper wheels, Dunlop racing tyres, special seats (a reclining one for the passenger) and a cranked gear-lever to improve its access from the reclined driving seat: the rev-counter was fitted behind the steering wheel, a smart wood-rimmed pattern on a lowered column. Rubber "Flik-switches" were fitted to the lighting and wiper switches.

A couple of problems presented themselves from our point of view—the racing tyres didn't agree terribly well with the road surface conditions of a soggy November, and from where we sat we could only see the rev-counter up to about 6,000 r.p.m. without leaning forward. The other snag was that the cranking of the gearlever meant that you couldn't select reverse gear until the handbrake was released—small points, but we mention them in case you're thinking of similar mods.

There was absolutely nothing, apart from the special wheels, about this car to suggest that its performance was in any way different from standard, which may be one of the reasons why the owner enjoys it so much! Since it is without doubt considerably faster than the majority of cars of up to 2½-litres the odd enterprising youth in a standard 1275 or other received a series of ugly shocks as 5 NOB disappeared into the dusk like a scalded cat. After all, 0-90 in 21 seconds isn't the sort of thing you come across every day, now is it? For all this, though, the whole thing was a vast improvement on the standard 1275 which, in

NOB "TAKES
FIVE" AFTER
A RUN UP
THE MOTORWAY
AT OVER 100

our opinion, falls short of the ideal—except in terms of performance and handling, which are both of the great fun order—in several respects. The driving position of 5 NOB presented one or two problems to us personally—obviously we aren't the same size or shape as Mr. Perrey—but as far as it went we felt that we could have sorted it out to be extremely comfortable. Certainly the seats were real seats, rather than perches. The engine was always easy to start whether it was hot or cold, and the tickover was steady and reliable at only a few hundred r.p.m. above normal—about eleven hundred revs, in fact. Flexibility was extremely good: driving in traffic could be approached without the need for lower gears or clutch slipping, and if the mood took you, you could potter along at under 30 m.p.h. in top gear— below two thousand revs—and accelerate up to a more appropriate lick without judder, snatch or any of the other road-test words which mean unpleasantness. The power real starts to arrive at about 3,000 r.p.m. and from there on

up it just keeps on coming. The standard car has markings on its speedometer to remind you to change gear at about 6,000 r.p.m., and one of the nasty features is the way everything roughens up for the last five m.p.h. or so in every case. But with the Broadspeed version this just wasn't so—when the rev-counter needle passes six thousand there's still a long way for it to go, and in fact you can go up to eight without any fuss. Obviously the engine is turning over pretty rapidly at that point, and it's only fair to allow it to feel like it—but it's still all properly smooth—you can cruise at over six thousand without any qualms at all, and a motorway speed of around the ton simply makes the water temperature go up a couple of degrees—nothing else happens at all, except that the motorway gets a lot shorter!

This is one of those cars where one is impressed less by what it does than in the way it does it. Top gear performance is all you could expect of a much larger engine, and the acceleration—in top—from around 70 up to three-figure speeds is pretty startling the first time you try it.

It goes without saying that good old Mini roadholding is well up to the extra performance. Quite how much the Cooper wheels and racing tyres contribute isn't easy to say—let's just agree that the thing handles beautifully under all conditions, with excellent steering, brakes and cornering. The only thing is that the racing tyres, although wizard for racing, as you might say, aren't entirely a good thing on greasy November road surfaces. And we have never realised until we drove this "dry" Mini what an improvement in ride with Hydrolastic bit gives you—although there again some of the bad riding habits are probably due to the racing tyres, which weren't constructed with riding comfort in mind!

There is, as we said, power from 3,000 r.p.m. upwards, and the real urge reports for duty at about five thousand. If you can keep the speedo needle above this you're really going places, and under these conditions the Broadspeed 1275 is not a car for watching birds—of any sort—in. With the sort of average speeds you can put up in this little projectile it is a Good Thing to stay on the ball, or disaster may well threaten. Which would be nasty.

Apart from the owners 19,000 miles, we added another few

Cooper wheels and Dunlop racing tyres.

YOU OPEN THE BONNET EXPECTING TO FIND SOMETHING PRETTY SPECIAL, WHAT YOU DO FIND LOOKS ALMOST STANDARD BUT IS INDEED VERY SPECIAL

hundred in the three days we had the car, and the thing never missed a beat.

We worked out a few figures just for laughs, and the overall average speed for the time the car was actually being used is something over 45 m.p.h.—including every kind of traffic condition and stops for fuel. Which means that it was going quite quickly for quite a lot of the time. Despite this we got an overall oil consumption of 200 miles per pint, as we said, and a fuel consumption of, overall, 24 m.p.g.—including the performance testing and whatnot. In other words, the converted car uses a lot less oil and only fractionally more fuel than the standard version, despite the fact that it chops more than three seconds off the 0-60 acceleration time, and adds about twelve per cent to the maximum speed. In slightly less than ideal conditions, we managed to get the car to accelerate from rest to 60 in 8·5 seconds, to 70 in 10·4, to 80 in 14·4 and to 90 in 21 dead: that sort of thing ain't peanuts, and it is clearly obvious that your hundred pounds is buying you a considerable increase in horse-

power over the 75 which the big Mini offers in standard form. The new maximum speed, taken as a mean of two ways, comes to 112·5 m.p.h., which is a lot of revs but, on the other hand, the power unit seems perfectly happy. Certainly it is perfectly satisfied with life at around the ton, which you can regard as a pretty rapid cruising speed on give and take roads—it seems to be asking a bit much of any small engine to sustain this sort of thing for sixty or seventy miles on a motorway. But for bursts of ten miles at a time it didn't bother it at all.

One way and another, it seems that the hundred quid asked by Broadspeed is well worth spending. You get a car which is faster than standard and, at the same time, is a damn sight more pleasant to drive or ride in. The Cooper "S" is no more a "mini" car, in the sense that this is a cheap economical runabout, than is a real live sports car (whatever that may be). After all, it costs the better part of eight hundred in basic form, so why spoil the ship for a ha'porth of power?.

 Cars on Test

BROADSPEED MINI-COOPER "S" 1275

Engine: Modified cylinder head, with larger inlet valves, ports and tracts; polished ports, recontoured combustion chambers; compression ratio 10·5 to 1. Modified inlet manifold and 1½ in. S.U. H4 carburettors; "road/race" camshaft; special exhaust manifold.

Transmission: As standard Cooper S.

Suspension: As standard Cooper S, with Cooper mag. alloy wheels and Dunlop racing tyres.

Brakes: As standard Cooper S.

Dimensions: As standard Cooper S.

PERFORMANCE

	m.p.h.			secs.
MAXIMUM SPEED	113·0	ACCELERATION	0–30—	3·5
Mean of two ways	112·5		0–40—	4·8
			0–50—	6·1
			0–60—	8·5
			0–70—	10·4
			0–80—	14·4
			0–90—	21·0

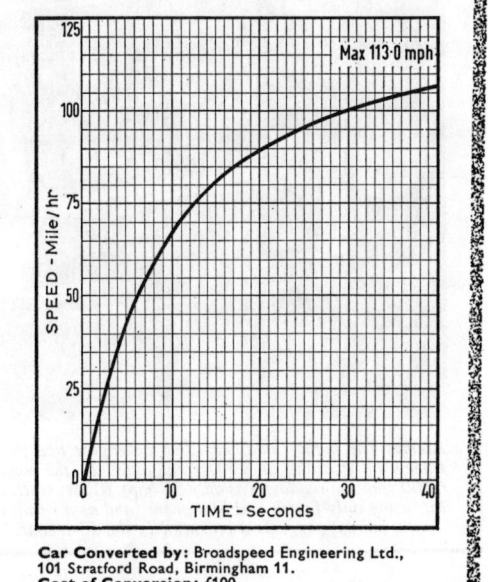

Car Converted by: Broadspeed Engineering Ltd., 101 Stratford Road, Birmingham 11.
Cost of Conversion: £100.

YOU could call it getting the low-down on Mini racing, because they just don't come any lower than this one. It is the latest 'baby' of Neville Trickett, a much-travelled and highly intelligent young man, who happens to be an accomplished painter, linguist and 'engine and chassis man', as well as a Mini-dicer of no mean repute.

During his three years of racing he has been in the forefront of the movement to extract unbelievable performance from the 'anything goes' type of Mini, and is a specialist in relieving them of excess weight. This latest car explores a new avenue—cutting down wind resistance—and the result is fascinating.

The intention had been to run this car in touring car races, but the day it was finished the 1966 regs appeared, stipulating standard bodywork, which promptly put the low-line Mini into the GT category, where it will be appearing this year.

ONLY 42 INCHES HIGH

The good looks of the converted car gave it obvious sales potential as a road vehicle, and a fully civilised version is being produced and marketed by GT Equipment Ltd, in which Neville Trickett is in partnership with Geoffrey Thomas. The Minisprint (road car) conversion, takes inches out of the height of a standard Mini, but is still a tall car alongside the racer, which is a bare 42 inches high, compared with the 53 inches of the normal Mini-Cooper.

The inches have been saved in several ways. First, 1½ inches have been cut out of the bottom of the side panels and doors, which have then been re-welded to the floor channels. A further 2¼ inches have been removed from the window pillars and roof supports, and the roof pan itself has been cut around its perimeter, so that

there is a more pronounced inwards slope of all the pillars and panels above the waistline. The front screen, for example, is at an angle of 45 degrees. Appropriate cut-outs have been made at front and rear, the frontal modification having the effect of bringing the headlights lower in relation to the grille opening.

In true club racing style, the interior of the car is completely gutted, with merely a pair of lightweight bucket seats and a stout roll bar added to the bare shell. Further lightening has been achieved by using glass-fibre engine and boot lids, ultra-lightweight doors of asbestos matting, and Perspex for all windows.

The low body silhouette tends to exaggerate the width of the track; in fact, the front wheels are 3 inches further apart than normal, and the rear track 5 inches above normal. The Minilite magnesium alloy wheels have 6 inch rims and carry Dunlop R7 tyres. Front suspension is normal competition Mini, with stiffened units and additional Koni shock absorbers, but the rear suspension is completely changed.

An unequal-length wishbone system has been adapted from modified Mini front-suspension links and mounted on to a special subframe—a very clever piece of work, this, because of course space is extremely limited in this area. Again, Konis augment the suspension units, and are mounted just outboard of them. Wheel toe-in is adjustable all round, while the front wheels run with the unusually pronounced camber of 4 degrees.

All the suspension modifications have been designed and carried out by Neville Trickett, as indeed have the modifications to the power unit, which started off as a normal 1,071 cc Mini-Cooper 'S' engine. The normal 40 thou bore-out has brought the capacity up to 1,098 cc, but later on special 1,145 cc pistons will be

fitted to bring the car close to the 1,150 cc class limit.

With a re-worked cylinder head and spe[cial] pistons the compression ratio has been brou[ght] up to 13 to 1, and there is a certain amoun[t of] originality in the modified breathing arran[ge]ments. The inlet manifolds (made up [by] Janspeed to drawings supplied by Trickett) [are] unusually short, with a rapid taper from 1⅜ [to] 1 3/8 inches, and a pair of SU HS6 (1¾ in[ch]) carburettors are used, which might have b[een] thought over-large for this size of power u[nit]. The exhaust system is also ultra-short, and [is] about 4 feet overall, including the 18 i[nch] megaphone at the end.

TITANIUM AND NEEDLE ROLLER[S]

The valve gear is lightened Mini-Coo[per] 'S' equipment with double springs giving a[p]proximately 190 pounds pressure. There i[s a] solid rocker shaft, and the rocker arms ha[ve] free-running needle-roller bearings, while [the] top caps are in Titanium alloy. Tricke[tt's] own camshaft, which also runs in needle rolle[rs,] has a timing of 55, 80, 80 and 55 degrees.

The basically 'S' pistons are lightened a[nd] the rods are end-balanced, while the cranksh[aft] is standard apart from very careful balanci[ng]. The competition diaphragm clutch a[nd] specially made thin steel flywheel have provid[ed] a further saving of 10½ pounds in weight.

A magnesium sump casing encloses a Mar[k] Colotti Type 40 five-speed gearbox, with mo[di]fied dog rings and steel selector forks. The fin[al] drive gears give a ratio of 4.78 to 1, and bo[th] these and the transfer gears are straight-c[ut]. Gear ratios are 1.0, 1.14, 1.35, 1.7 and 2.7 to [1].

At first glance, Neville Trickett's car loo[ks] undrivable—you just can't get down that lo[w.] Of course, you soon find you can, and once

Above: *The highly modified 1,098 cc engine has shown a remarkable 98 bhp at the wheels on a brake at 8,500 rpm, albeit at the expense of middle-range torque (at 6,000 rpm the reading was only 55 bhp).* **Right:** *Getting down to it! A lap in 62-and-a-bit, using only fourth and fifth gears (and as a result being well off the cam at several points) suggests a potential in the 58-second bracket at Brands Hatch.*

GT Equipment's 'low-line' M[ini]

inside the box I was embarrassed not by ⟨lack⟩ of room, but by an excess of it. Single-seat drivers can't teach these Mini boys a ⟨thing⟩ about lying down to it, and with the seat ⟨welded⟩ in place to suit Trickett, who is some ⟨inches⟩ shorter than myself, I could scarcely ⟨reach⟩ the steering wheel! Someone found a ⟨thin⟩ coat to slip between me and the seat ⟨backrest⟩ (I use the last word in its loosest sense!) ⟨and⟩ although I couldn't claim to be really ⟨comfortable⟩, at least this put me within reach ⟨of all⟩ the bits that mattered.

⟨There⟩ is quite a routine involved in getting ⟨into⟩ first or reverse. The normal gear stick is ⟨augmented⟩ by a parallel lever rather like a ⟨shooting⟩ stick, which has to be twisted to the ⟨right⟩ before either of these gears can be reached. ⟨This⟩ 'shooting stick' springs back to the right ⟨immediately⟩ any of the other gears are selected, ⟨after⟩ which first and reverse are unobtainable, ⟨the⟩ five-speed box becoming in effect a normal ⟨four⟩-speed unit.

⟨There⟩ is a heavy spring bias towards second ⟨and⟩ third, so that quite a firm pressure is ⟨needed⟩ from the left side of the stick when ⟨selecting⟩ fourth and fifth. This didn't trouble me at ⟨all⟩, but third gear proved completely elusive, ⟨so⟩ much so that I carried out the test for the ⟨most⟩ part in fourth and fifth.

⟨Although⟩ this did not prevent an all-round ⟨appraisal⟩ of the car, it did, of course, have its ⟨effect⟩ on lap times, the more so because this ⟨power⟩ unit is unusually 'top-endy'. On a brake, ⟨recently⟩, as much as 98 bhp was seen at the ⟨wheels⟩ at 8,500 rpm, at the expense of flexi-⟨bilit⟩y, the figure at 6,000 rpm being a modest ⟨⟩ bhp.

⟨B⟩eing a very new unit, I adhered to a rev ⟨limi⟩t of 8,500 rpm, which was a mere 700 rpm ⟨abo⟩ve the point where there was a noticeable build-up of power (I would imagine there is quite a 'lump' in the power curve at 7,800 rpm). This meant, of course, that I was 'on the cam' for perhaps only two-thirds of the lap, and was very slow out of the hairpin, and a best lap time of between 62 and 63 seconds suggests a 58-second potential by the car (which, of course, will be needed in the highly competitive GT category).

For a first-time out (the car had not previously been driven on a circuit) the Mini handled exceptionally well. This was my first experience with the world of ultra-wide-track Minis, and the improvement in stability by this and by the generous rubber contact area is highly educational! The understeer I found by no means excessive (although Neville Trickett himself said he was used to less), although I did notice the fairly marked alteration in attitude between power-on and power-off. Only in these terms, I think, would I have found less understeer beneficial.

TWO-WAY ADVANTAGE

The advantage of a low body line in reduced wind resistance is obvious, but the psychological advantage of such a low centre of gravity must also be worth something—it felt as though you would need the assistance of a crane to roll this one! The inevitable choppiness of the ride calls for firm support for the driver if he is to achieve maximum effort, and the owner was right to select his own ideal position and then weld everything up . . . even though it didn't happen to suit me!

The brakes were suffering from newness, but though they lacked ultimate stopping power they seemed to be in good balance, and pulled the car up square every time.

Some hesitation in pick-up was probably caused by dwindling fuel supplies—the aluminium tank holds only 3 gallons, which does not go far with a thirst of 8 or 10 mpg . . . yes, this one really drinks it! (Indeed, it drank the lot on the last lap!)

Neville Trickett and GT Equipment seem to be on to a good thing with their low-line Minis. It needs something as bold as this to breathe new interest into the Mini class, for there is a limit to the attraction of a load of cars, which at least *look* identical, racing against each other.

New regs have pushed this car from the Touring to the GT category, where it will be up against things like the Mini-Marcos and of course the Diva. But it needn't be outclassed, for all that. An all-up weight of about $8\frac{1}{2}$ cwt, minimal frontal area, and close on 100 bhp at the wheels (when you can match your gearing to the high and narrow rev band) plus best Mini handling adds up to an impressive lap time. You wait and see!

⟨Ra⟩cing Mini

TRACK TEST

No 48 by John Blunsden

'If you think that one's low, you should see the one they lent me!' Stirling Moss tried the Minisprint road car while Blunsden played with the starker version. Like it? He's having his own 'S' converted!

Just the bare facts. Revs, oil pressure and water temperature are all you need to know. Long gear stick is connected to a Type 40 Colotti five-speed gearbox. A second 'umbrella handle', hidden behind the steering column, has to be twisted to allow engagement of first or reverse. Despite low roof line and driving position, visibility is entirely adequate.

IT wasn't meant to be fun, quite. We were setting out to test a sumpful of Duckham's Q 20-50. The exercise was one of a series of tests under different conditions. With this one we were looking for extremes. The Sahara would give us very hot and very cold weather, and would work the engine hard for unusually long periods. Poor fuel would dilute the oil. Dust and sand would penetrate everything, and what had worked well in racing tests and survived hard running on the M1 might not work so well in the desert—especially in a Mini, where engine and transmission share the same oil.

Our standard 1965 998 cc Morris-Cooper would clearly need a lot of preparation. Minis were never designed to cross the Sahara and a failure could mean disaster. Various firms—Smiths, S.U., Dunlop and Firestone—provided essential equipment. Kenlowe fitted what was probably the most elaborate cooling system every put into a Mini; but most of the work was done by University Motors at Hanwell, who checked and re-checked everything on the car, fitted a mass of extra equipment and delivered to us the best equipped Mini in the world for desert crossings.

Through France, Spain and Morocco, we had rain, gales and snow, but on our tenth day we set out from Bechar in Algeria into the desert. We were carrying ten gallons of water and twenty gallons of petrol. The load on the Mini was nearly three-quarters of its own weight. Although we were still on metalled roads, within 60 miles of Bechar, we hit our first real obstacle. Instead of a road there was a fast-flowing river 60 yards across. There had been heavy rain on the High Atlas and a causeway had been carried away. There was a ford of sorts, but it was 2 ft. deep and very rocky. Numerous French army lorries were trying to make a new ford, but their drivers said they had no authority to carry civilian traffic. After a few hours we got a lift from a local lorry and carried on. We were now in the desert. Great sand-dunes frowned down on the road, still metalled; wrecked cars lay still beside it.

The road went on through a gap in the dunes and straight into another river, but this was easily forded. Suddenly, at dusk, the road stopped. Diversion. Oil drums across the road and an arrow pointing vaguely at the desert. We turned off rather hesitantly, missed the correct track, and were promptly stuck in some sand. Deep, loose, yellow sand. We tried our sand mats. They were too flexible and didn't work. The car sank right in. As we puzzled what to do the first vehicle for hours appeared—a lorry laden with Arabs. They pushed us out, and the driver said gaily to follow him, as the "piste" was easy to lose. He set off; we reloaded the car and tried to catch him up. This was our first piste; it was old and disused and the corrugations were bad with many rocks. We caught our Arab guide up, but on a particularly

MINI ACROSS THE SAHARA

In November last year, Tim Miller and Jonathan Green-Armytage left London in a Mini-Cooper to carry out oil tests on Duckham's Q 20-50 in the Sahara. The full story of their adventures has already been published in a small booklet produced by Duckham's, but it was felt that readers would be interested to know how the Mini can stand up to such extreme conditions

BY T. MILLER

bad bump the rubber petrol tanks which we had lashed so carefully on the bonnet came adrift, and one burst as we ran it over. We picked up the other and continued more slowly, but still the noise and vibration were frightful. We stopped for the night at a mysterious unsignposted fork in the piste, cooked a meal, and slept out. It had been quite hot that day, yet during the night the temperature dropped almost to freezing.

Next morning we checked the car and found that the horn had broken loose from its mounting, and the roof rack had almost collapsed. Otherwise it was fine. The puzzling fork in the piste meant nothing—the two roads joined up in a couple of hundred yards. But this section was worse than ever. It was very rocky and had some patches of deep sand that had to be taken at speed. One of these sent sand flying right over the windscreen and over the car like a big puddle. At the bottom of it there was a rock which cut a deep groove in the sump guard. Eventually we came to the end of this diversion and came back on to the metalled road which we knew would end in a few miles. We checked the car again. The roof rack had finally collapsed and had to go. So did the useless sand mats. The spare wheels went on the bonnet and, with some cunning packing, everything else fitted inside the car. We carried on, gave some rubber solution to a lorry-driver stuck with a puncture, and came to the formal end of the surfaced road. From now on we should be on piste.

One imagines Sahara "roads" being endless, smooth, flat sand, but instead they were hard, like corrugated iron, with the corrugations about 2 ft apart. Sometimes this is strewn with rocks, sometimes covered with deep rutted sand, but mostly the piste is one long, vicious, drumming bump. The steering wheel twitches madly and, until the spokes of our non-standard wheel were bound with insulating tape, fingers were being cut to shreds by the metal spokes. Tiny wheels, high-pressure suspension and radial tyres make it worse, but we needed the high pressures in the Hydrolastic suspension to give us every inch of ground clearance we could get. Very often the desert beside the piste was much smoother than the piste itself, but we were reluctant to use it for fear of breaking through the crusty surface into soft sand and having to wait a couple of days for the next lorry to come along. We slept out that night, and again it was cold.

Next day we passed through Adrar, a big oasis and administrative centre, and Reggane, the French satellite base. It was hard, rough, deserted piste all the way. The only respite was a stop at Adrar to film a camel caravan at a well, and to buy petrol and have lunch. After Reggane, the piste became sandy. Not deep—usually only a couple of inches—but there was always the threat of a deeper section, and the speed had to be kept up. There was one big problem. The piste kept dividing itself into two, three or even four. We couldn't slow down. It meant instant decision, with the car swishing over sand at 60 m.p.h. with only headlights to help in the choice. At the same time the car was sliding about like a skier. That was the time when the Mini's virtues were most apparent. With front-wheel drive and precise steering it could actually be held on a line. After rough piste this was exhilarating, but it needed concentration. Eventually we took the wrong line, were stuck and fell out of the car straight into our sleeping bags.

The next day, after a difficult start from the sand, we reached Ain Salah where we checked the oil and changed its filter. Only two pints of oil had been used in 3,167 miles. The heat was not affecting consumption. The local lorry drivers had never seen a car so small, and they doubted our ability to reach Tamanrasset. On the 15th day we set out on the middle crossing. Destination: Tamanrasset—the centre point of the Sahara—432 miles of nothing. No petrol, little water. The car was loaded up with 20 gallons of fuel and we drove out through the soft sand at full revs. in third. Two miles outside Ain Salah we got the route wrong and were stuck on a disused piste. It was a hot day. We had to dig, and dig, and dig. It took four hours to back out a hundred yards. Our salvation was the hard surface under the sand; but the sand was a foot deep.

Under the Stars

Even when we found the right piste, it was a sandy one; but swishing along at a good speed we had no trouble until the sand stopped and we felt steep corrugations, jolting everything again. We stopped at the first excuse that night, beside a great yellow aviation spirit tanker that had passed half a mile away when we were stuck in the sand. We ate with the lorry crew and slept under an amazing sky full of stars. It was the idyllic Sahara.

On the 16th day we started before dawn, and drove on and on over ordinary piste, rocky piste, sandy piste. We refilled the main tanks from the auxiliaries, and later passed the great black mountain where the French keep their atom bombs, and thought of stopping for the night. We stopped beside some lorries, but learned that a convoy was going south from Tamanrasset the next day, so we pressed on until, 120 kilometres short of Tamanrasset, the left rear suspension unit burst. We drove on very slowly for 10 kilometres, and stopped for the night. ▶

Left: Early stages—before the roof rack collapsed and had to be abandoned, along with the sand mats. Below: Through a fly-spattered windscreen—a typical view of the lonely, cairn-marked, desert track

MINI ACROSS THE SAHARA

Next morning we confirmed that it was the unit itself, and not a burst pipe. The handbook said that if the suspension failed the car could safely be driven at up to 30 m.p.h. on metalled roads to the nearest B.M.C. distributor. Our nearest B.M.C. distributor was in Kano (1,088 miles) or in Algiers (1,194 miles). The nearest metalled road was 650 miles away. We decided to see if we could at least reach Tamanrasset.

The piste to Tamanrasset was quite good piste, as piste goes, but because our left-hand ground clearance was now reduced to two inches, we made slow, agonising progress, and ticked off the marker posts that appeared every 10 kilometres as though they were days left in prison. We passed the Tropic of Cancer signpost; and at last we arrived.

We had travelled 3,618 miles in a running time of 104 hours. We had used 126 gallons of fuel and a mere two pints of oil. The suspension had burst because we were carrying a heavier load than that which the Mini was intended to carry; over tracks for which the Mini was never designed; and with Hydrolastic pressures 50 per cent above normal. Everything else on the car was fine. The engine was amazingly crisp and there was no play in the steering. There were three possibilities left for us. We could repair the car, which would mean going to Algiers to get a spare and would take at least a fortnight, assuming there was one there; and there was no pressure pump in Tamanarasset. We could put the car on a lorry and take it south to Kano, where spares would be fairly easy to get; but the Niger border was closed to lorries—had been for a month—and no one had any idea when it would be open. Or we could put it on a lorry and take it back north. This was the only realistic thing to do. We found our lorry on our fourth day in Tamantasset. We ran out of cigarettes on the third, and the nearest shop was 300 miles away. That was the thing that made us feel most remote. The lorry took us and our Mini, along with assorted Arabs, Touareg, firewood and empty beer bottles, 1,200 miles to Algiers. We either froze at night on the back of the lorry, or suffered cramp in the cab. The car bounced about and became badly scarred by the ropes that bound it down. We suffered, and the car suffered, but that lorry took us to Algiers. On Christmas Day we sailed with the car for Marseilles—we had not been able to find a spare in Algiers. We travelled fourth class, and there was a storm. We had nothing but sardines for our Christmas dinner. But we reached Marseilles and had the car in action again in half a day. ∎

Top: For running repairs and service, natural gullies make an improvised ramp.
Above: Discarded lorry tyres, dumped in the desert. Below: "We had to dig, and dig, and dig. It took four hours to back out 100 yards"

JOHN BOLSTER tries

A works Mini-Cooper S

HARRY KÄLLSTROM and Ragnvald Hakansson speed the Mini through Sherwood Forest during the RAC Rally.

THE performance of the BMC Minis in rallies has been one of the most spectacular developments of recent years. I am not a rally man myself, but I am greatly interested in the mechanical side of any successful high-performance car. Accordingly, I arranged to borrow the highest-placed Works Mini in the RAC Rally, the idea being to get a representative one, exactly as it finished.

The results of the RAC Rally are too well known to require further discussion, and naturally it was the car of Källstrom and Hakansson that I took over. The machine is an Austin-Cooper 1275S to Appendix J Group 2 specification. The engine is overbored 0020 in, giving a capacity of 1293 cc, and has a compression ratio of 11.4 to 1. The cylinder head is the work of Daniel Richmond; a standard camshaft is used and the twin SU carburetters are of 1½ inch bore. There is a competition exhaust system with a modified exhaust pipe. The competition clutch is fitted to a lightened steel flywheel, and the gearbox has close-ratio straight-cut gears; a top gear ratio of 4.26 to 1 has been chosen.

As is normal with modern rally cars, an immense amount of special equipment is fitted, the extra current consumption being looked after by a Lucas alternator. An extremely rugged full-length steel undershield is employed and this is most necessary, as the car bottoms continuously on rough roads, such as the drive up to my house. All pipes and electric cables are carried inside the coachwork to prevent damage, and a roll-over bar is fitted. As tested, the car was shod with Dunlop R7 racing tyres at very high pressures, exactly as it had run in the Silverstone test.

Having driven the car hard for several days, I can say that it finished the Rally in excellent condition. Its only peculiarity was an occasional tendency to pull to the left, a legacy of the hard use to which it had been put, no doubt. Apart from this, the steering was precise, the brakes powerful and even, and there were no suspicious noises.

The suspension is hard and the car rides very flat, being utterly controllable at all times. It is very well balanced, with no tendency to lose either end unless the driver wants it that way. The most astonishing feature is the tremendous punch of the engine at low revolutions, which is a Daniel Richmond speciality.

It is permissible to run up to 7200 rpm, at which revs the car reaches about 38 mph on first speed, 55 mph on second, and 78 mph on third. On top, using the same maximum revolutions, a genuine 100 mph can just about be touched, and very quickly, too.

With all its equipment, the car is by no means light, but the acceleration is really fierce. As so often happens in November, the roads were wet throughout the period of my test, so it was pointless to make a graph of the acceleration. On a wet road,

I was able to go from a standstill to 60 mph in 9 seconds, so one can easily guess at the sort of figure that would be possible under dry conditions.

The clutch gripped instantly at all times, tending to be a little fierce. The gearchange of this car is the best I have yet handled on a Mini, first speed being curiously easy to engage. The exhaust is well silenced and the engine is so flexible that traffic driving is a pleasure. I used the Cooper in London without any tendency to oil plugs, though the engine must be set to idle fast to avoid stalling. Above 4000 rpm a sort of boom is produced inside the car which renders conversation difficult, but there are no unpleasant mechanical noises, most of the disturbance coming from the unsilenced carburetter intakes. The whine of the transmission can be heard from the front of the car outside, but it is not obtrusive to the occupants.

There is no doubt that this particular Mini has finished a gruelling rally in perfect health. It is the greatest possible fun to drive, the proper bucket seat giving perfect location, and the passenger can turn a handle if he wants a reclining position for resting. In a word, this is a **tough** little car.

Cibié introduce dipping iodine headlamp unit

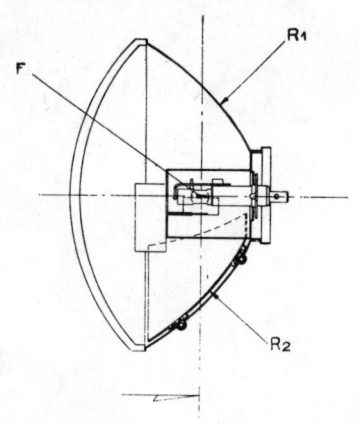

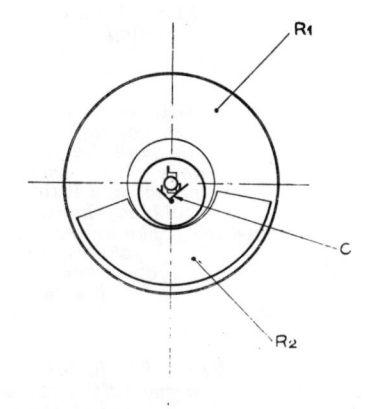

AT the end of January 1967 a revolutionary headlamp unit will be available from Cibié; it makes it possible to utilize an iodine filament dipping system in conjunction with the main beam. Known as the DI7 (dipping iodine 7 ins), the unit is designed to replace existing normal headlamp bulb systems on cars where it is not practical to adopt the recommended fourlight layout, as fitted on several makes such as Rover, Alfa Romeo, Fiat and so on.

Briefly, the Cibié DI7 employs a singlefilament iodine vapour bulb; to achieve dipping, a pair of hinged shields cuts off the light reaching the lower area of the reflector. A glance at the accompanying diagram will show how this is achieved. The unit comprises a parabolic reflector (R1) in the lower portion of which a small reflector (R2) is located; this is a segment of the parabola.

The lamp filament is situated in front of the focal point of reflector R1, from which the dipped beam is produced. To make certain that only the beam emanating from the upper part of the reflector is used, a pair of shutters (C) are provided below the

iodine bulb; a solenoid is used to cause them to lie vertically below the bulb, thus producing the main beam with full reflector area. Operation of the dipswitch returns the shutters to the masking position. The unit itself has been so designed that, in the unlikely event of mechanical failure, the lamp will "fail-safe" and remain on the dipped beam.

The assembly has been subjected to lengthy reliability tests, including vibrations which are greatly in excess of those ever likely to be experienced on road cars. Great attention has been paid to excluding road

grit and damp from the unit, and the bulb itself is enclosed in a glass shroud.

Thus, one can now have powerful iodine lighting with instantaneous dipping, which has the advantage of reducing the fractional period of transfer from bright to dim, the latter giving an exceptionally good spread without dazzle effect whatsoever.

For the present, the units will be obtainable with either l/h or r/h dipping. Including taxes, prices will be £17 17s 6d the pair, including bulbs. Bulb life is calculated to have a minimum of 200 hours, and standard replacements are 28s 6d each.

Now, more than ever, BMC's wonderbox has become the Super-Cooper.

By ROB LUCK

BECAUSE SPORTS CAR WORLD staffmen spend a great deal of time climbing in and out of motor cars of all descriptions we tend to consider ourselves immune to the surprise factor in road testing. Indeed, looking back over the past 12 months it is fairly safe to say that very few motor cars have surprised us in any way. There is one notable exception — the Morris Cooper 1275 S type.

The car could never be classified as a sports car on any basis, yet the argument which has been used too fondly by some writers in supporting the case of pretenders to the title of "sports car" probably applies to this car more than any other: it does most things that a sports car is noted for, but in a far superior manner. The Morris Cooper S has never pretended to be a sports car (and any such claim would be laughable). Instead, its tiny dimensions fill a huge gap in the market without really leaving much room for any other marque. It is essentially a sport-

SPORTS CAR WORLD · ROAD TEST

For road and track the
1275 Mini Cooper S is ... **THE**

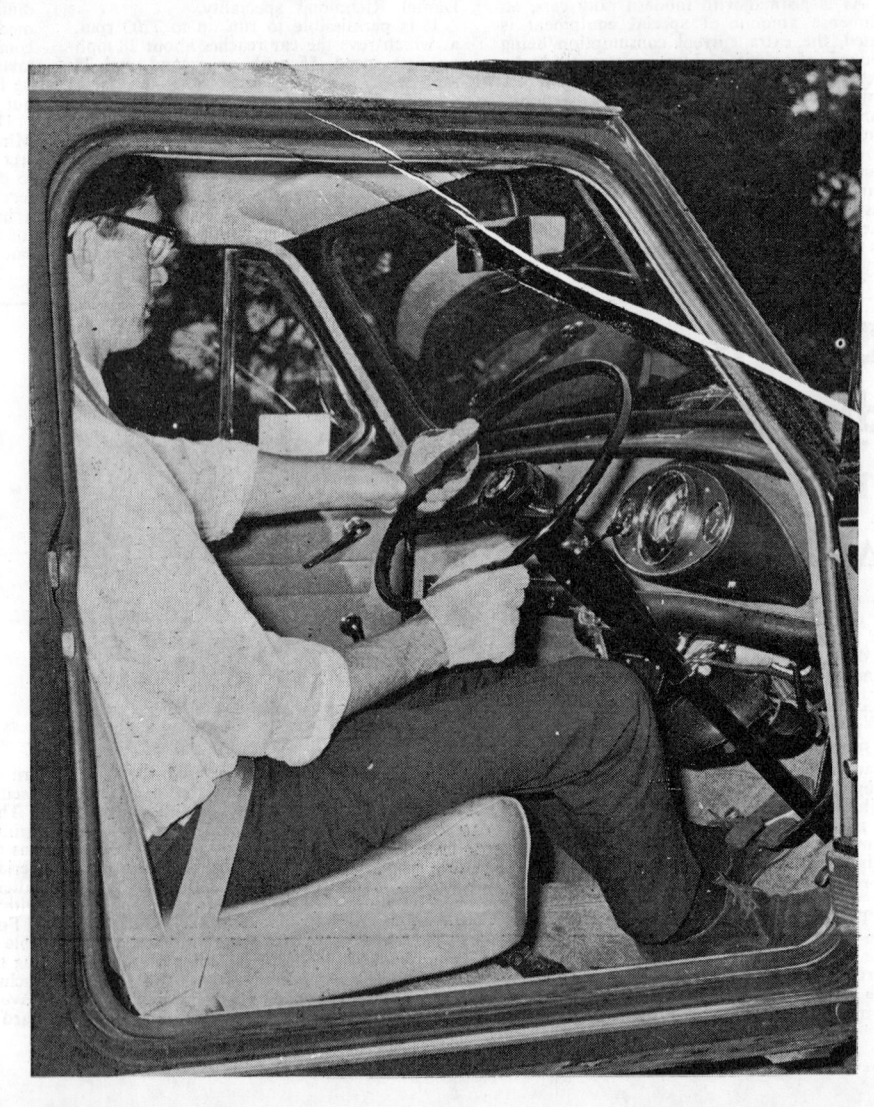

Going away: you might be lucky to pick up the tiny S clue to why you lost out in a back-street match race. Twin fuel fillers drain into single feed.

Driving position is comfortable through the full range of boy-racer crouch and little-old-lady hunch. The driver is always in command of his car.

The LSD factor— "slippy diffs" have given a new meaning to Mini handling: here Brian Foley pours on power and lock in a 105 bhp Cooper at Warwick Farm.

ing sedan — nothing else; but it responds equally to the roles of sports car, Grand Tourer, racing sedan, city run-about, family hack, second car and wife's transport.

There are those who would discredit the car for its almost miniature construction but in this regard the car is a complete paradox: it has a far broader design concept than many twice its size. The car's tiny dimensions (10 ft long by 4 ft 7 ins. wide by 4 ft 5 ins. high) have not affected its climb to the position of the most versatile competitor in motor sport.

The test car emphasised BMC Australia's emergence as an autonomous body, independent of the English parent company. It was the first Cooper in the world to combine Hydrolastic suspension and wind-up windows with the enlarged A series power plant. This and the lowly price of just over $2200 (£1100) came at a time when our English counterparts were screaming for any 1275 S types they could get their hands on. Since its introduction it has been adopted by South Africa, considered by England and just as by-the-way become the most popular high performance sedan in Australia.

The car, like any other, is not faultless, but you will search long and hard — and fruitlessly — for a car which offers performance, safety, reasonable comfort and economy all for the tiny price tag it carries. It is this combination of offerings which causes most surprise. For a motor race and rally winner you could justifiably expect a rough, lumpy-engined, noisy, brisk-riding and temperamental motor vehicle. The Cooper S is none of these.

The Cooper strongly resembles the Mini DeLuxe in appearance. This has earmarked the car as a prime Q-ship and we still rather fancy the idea of stripping off the S badges and substituting DeLuxe badges for the total effect.

The car has every basic attribute of a high performance vehicle: it goes, stops and handles without vices although, allowing for the eccentricities of the great, it certainly has its peculiarities which require adjustment by the unfamiliar driver. Perhaps some who are very familiar are also aware of its unusual characteristics; no doubt Peter Manton will have something to say about committing the car to a line when something is blocking the path. But as with any other car there is a margin for error — the Cooper probably leaves a greater one than most.

Because of their unlikely size Minis have always been the favorite of Australian race crowds and recently this has increased with the advent of the Mustangs. Mini drivers have been a colorful lot and none who saw it could ever forget the monumental dice between Johnny Harvey and Ian

GIANT-KILLER

Geoghegan at Warwick Farm, or other epic battles at Catalina Park, Lakeside, Sandown and other Australian circuits. The car's International rally successes speak for themselves.

Perhaps the focal point of Minis has been their revolutionary handling and it is such a familiar subject with all Australian motorists that it needs no further explanation. Suffice it to say that its power on-understeer, power off-oversteer characteristics have such great natural handling attributes and built-in safety margin that the car is one of the most manoeuvrable ever produced. It is also one of the safest. When combined with such devices as limited slip differentials, wide rim wheels and so on the possibilities become unlimited.

Braking per front discs and rear drums is one of the car's better features. On the test car we obtained good figures but at the expense of some tricky moments when the right front brake locked up completely and wore flat spots on the tyre. As this was completely beyond our previous experience with Minis we re-traced the car's history and among its list of activities found that it had been raced the day before our test by another energetic tester. The badly worn tyres had not been replaced — an error which the BMC Public Relations department is thankfully not in the habit of making. The car was otherwise well prepared and proved a very inviting test unit decked out in BRG with white top.

Inside, the green decor ran to full carpeting on the floor and BMC's new heavy duty vinyl trim for seats and door lining. The instruments are grouped together in an oval binnacle in the centre — obviously built up around the old central speedometer system that has been the feature of all Minis. The two supplementary gauges are for oil pressure and water temperature. The big speedo carries the fuel gauge and warning lights to cover all other functions. Under the binnacle is a switch to light all instruments. There is no tachometer — virtually an anachronism in such a car in the year 1966. BMC would have been well advised to provide this key to the free revving motor's behavior at the expense of adding to the low purchase price by $10 - $20. At least the gadget should be offered as an option.

A heater has been added as standard equipment and is operated by the strangest push-pull knob we have seen for a long time: push-in is on, pull-out is off. The system works well enough though and the car warms through fairly quickly.

The wheel location — famous for inducing the bus driver's crouch and other similar maladies — has probably done more than any other driving position to induce a near to straight-arms position in every driver from little old lady to boy-racer. The seats definitely seem better designed orthopedically than previous Minis and long hours at the wheel don't become nearly as tiring. The wheel itself is that same uninspired huge black plastic affair that relays unbelievable quantities of information to the driver's hands without transmitting road shock or vibration. It is a simple matter for even the inexperienced to know just how much traction is available at the front wheels at any time.

With such an essentially functional driver set-up the little lapses into mediocrity evident here and there become all the more inexcusable. A headlight flasher is badly needed and the dipper switch would best be incorporated in a stalk on the steering wheel rather than in the finickity hide-and-seek position on the floor. There should be a government edict on headlight penetration and windscreen wiper speed on high performance cars and the Mini in both these requirements lacks adequate equipment. But the car is fitted with windscreen washers which do a reasonable job, although mechanical.

Ride quality of the Cooper is one of its most unbelievable characteristics. There is such a perfect balance between the respective facets of the car's performance — acceleration, braking and handling — that it is impossible to catch it out of step, or off-balance. It simply goes where it is pointed, booted or stopped, with the least possible effort and without any discomfort to passengers. It will straddle huge potholes, charge over rough terrain or motor down the expressway with equal imperturbability. Twisting mountain roads, creek crossings are negotiated as a matter of course and in its unstoppable way the car does everything effortlessly. It sets up an uncomfortable fore-and-aft pitch on some bitumen roads, but this is never annoying.

One of the leading questions which is directed at small high performance cars is noise level — how great? The Mini Cooper S is not without its share of problems, but they are better described as the inadequacies you might expect in an underpriced high performance car, rather than major disadvantages of the vehicle design. Interior boom does not intrude at low to medium speeds, but at high average speeds it does become somewhat of a problem. Fortunately there is an extremely

The eyes have it. Minis' tiny headlamps need more powerful units to cope with sustained high speed night cruising. Spotlamps are easily fitted.

J. Harvey Esq. gets well and truly crossed in the Esses at Warwick Farm in his Austin version of the 1275 Cooper S. Car is fully imported.

MORRIS COOPER "S"
(1275 cc)

SPECIFICATIONS

DIMENSIONS:

Wheelbase	6 ft 8.5 in.
Track, front	3 ft 11.5 in.
Track, rear	3 ft 10.25 in.
Ground clearance	5.8 in.
Turning circle	31 ft 6 in.
Turns, lock to lock	2.33
Overall length	10 ft 0.25 in.
Overall width	4 ft 7.5 in.
Overall height	4 ft 5 in.

CHASSIS:

Steering type	rack and pinion
Brake type	disc front, drum rear, servo assisted
Swept area	177 sq in.
Suspension, front	independent by Hydrolastic displacers interconnected front to rear
Suspension, rear	independent by Hydrolastic displacers inconnected front to rear, competition rate Hydrolastic fluid
Tyre size	145 mm. 10 in.
Weight	13 cwt.
Fuel tank capacity	5¼ gals per tank (two tanks)
Approx. cruising range	300 miles

ENGINE:

Cylinders	four, in line
Bore and stroke	70.6 mm by 81.33 mm
Cubic capacity	1275 cc
Compression ratio	9.75 to 1
Fuel requirement	60/40 super and methyl benzine
Valves	pushrod, overhead
Maximum power	75 bhp at 5700 rpm
Maximum torque	79 ft/lbs at 3000 rpm

TRANSMISSION:

Overall ratios	
First	11.02 to 1
Second (synchro)	6.60 to 1
Third (synchro)	4.67 to 1
Fourth (synchro)	3.444 to 1
Final driven	3.444 to 1
Mph per 1000 rpm in top gear	16.1

PERFORMANCE

Top speed average		95.4 mph
Fastest run		96.1 mph
Maximum, first		37 mph
Maximum, second		59 mph
Maximum, third		84 mph
Maximum, fourth		96 mph
Standing quarter mile average		17.9 seconds
Fastest run		17.6 seconds
0 to 30 mph		3.7 seconds
0 to 40 mph		6.0 seconds
0 to 50 mph		8.2 seconds
0 to 60 mph		10.8 seconds
0 to 70 mph		14.9 seconds
0 to 80 mph		19.2 seconds
0 to 90 mph		NA
0 to 100 mph		NA
0 to 60 mph to 0		13.6 seconds
	Top	Third
40 to 60 mph	8.6 secs.	6.7 secs.
50 to 70 mph	10.1 secs.	7.8 secs.
60 to 80 mph	12.4 secs.	NA
Fuel consumption, overall		26 mpg
Fuel consumption, cruising		25-29 mpg

simple answer — one of the excellent noise deadening kits available at cheap price on the market should reduce the problem to insignificance.

In the Cooper S, BMC has lowered the price of performance to a reasonable level for the average buyer. Somehow it has maintained an amazing balance of power and economy — so that the car actually comes closer to meeting the requirements of a larger number of people. Any compromises that have to be made can be more than equalised by the advantages. The car is a low-purchase, low maintenance vehicle with tremendous built-in longevity and a durability exceeded by few far above its price. It is a revelation for many people driving more expensive cars.

The Cooper S type has already become a virtual legend in its own time — and not without reason, for it combines every desirable attribute in a high performance car. It can be driven straight from supermarket to race track without any fiddles. It does everything well and everything effortlessly. Perhaps the greatest compliment that stands as tribute to the car is that it needs no introduction in any circles. #

You have to be tough to take this Tulip!

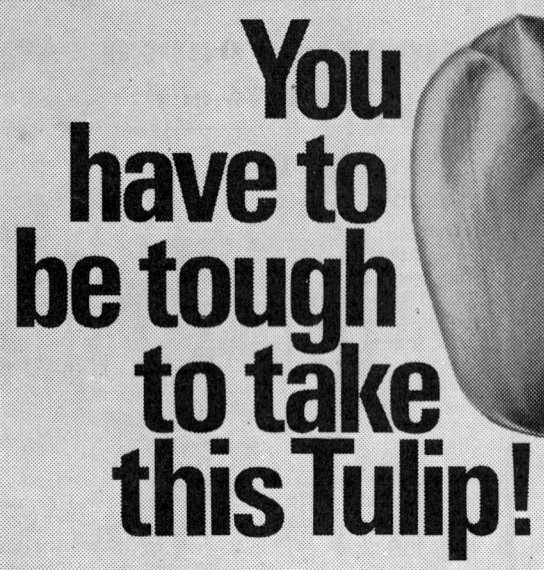

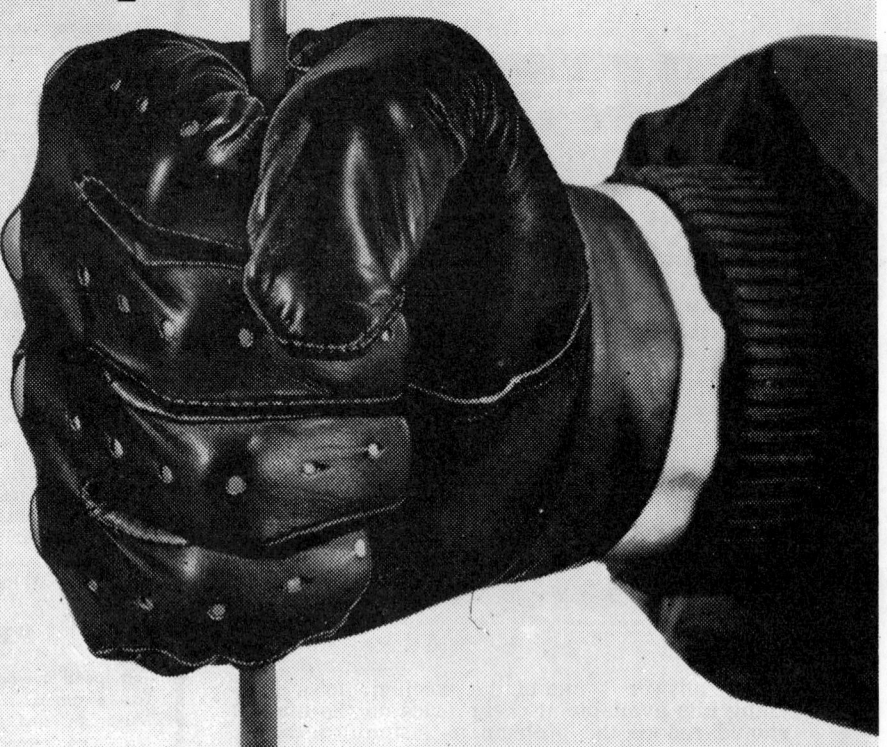

It wasn't tulips, tulips all the way for BMC in Holland's famous motor rally. In fact, it was as tough as rallies can get. This second consecutive victory proved just how much BMC engineering can take — and more. More than you'll ever give your BMC car, perhaps. But it's built to take lots of punishment. All BMC cars are.

Austin · Austin-Healey
MG · Morris · Riley
Vanden Plas · Wolseley

BMC

THE BRITISH MOTOR CORPORATION LIMITED
LONGBRIDGE, BIRMINGHAM

BMC SCOREBOARD: TULIP RALLY

BMC WIN TOURING CAR CATEGORY

1ST **MINI COOPER S**	2ND **MINI COOPER S**
Timo Makinen/Paul Easter	*Rauno Aaltonen/Henry Liddon*

ALSO MANUFACTURERS TEAM PRIZE

MINI COOPER S	*Timo Makinen/Paul Easter*
	Rauno Aaltonen/Henry Liddon
	Julien Vernaeve/Mike Wood

Subject to official confirmation   **108 STARTERS 65 FINISHERS**

TRACK TEST

■■■■■■■■■■■■■■■■■

JANSPEED 1293 COOPER S

Clive about to do the wall of death act with Geoff—he looks happy enough about it though!

THE day dawned wet and blustery, and I viewed with some apprehension the day which I had arranged to spend down at Castle Combe with the Janspeed Racing Equipe. However, by the time we had reached the circuit and put a bowl of soup, some sandwiches and a pint inside us the rain had left off and the sun was shining. A fairly strong wind quickly dried everything out.

The car I was to test was the latest Janspeed 1293 c.c. Cooper S, painted in the team's familiar colours of red with a black roof, aptly registered JAN 4.

This vehicle had been purchased in December 1965 as a rather scruffy but sound 1960 Mini; needless to say it was at that time mechanically clapped.

The original intention had been to rebuild the car as a really potent 850 Racer and this is why a 1960 car was purchased, as Minis of this vintage show a weight saving of about 40 lb. by virtue of using thinner gauge metal in their construction. It was felt that with only 850 c.c. to play with weight was of prime importance. These plans were in fact carried out and an 850 Mini (in fact only 825 c.c.) was constructed using all the normal Mini tuning extras such as Tec-Del Minilite Wheels etc. The engine was constructed around a 970 c.c. Cooper S crankshaft and an "ordinary" 850 block. A five-speed Colotti-Francis gearbox transmitted the power. However, with the lack of enthusiasm for 850 racing displayed by race organisers this season the project was never raced on a circuit, but just collected dust in the workshop.

With the sale of the Janspeed "works" Mini-Marcos Jan Oder had to look around for another works racer. Needless to say his eye fell on the 850.

Quickly the car was stripped down again and rebuilt as a Cooper S, albeit with a somewhat lighter body, which cannot be a disadvantage.

This brings us up to date, when I tested the car.

For the technically minded the specification followed fairly "standard" Mini practice.

Although the body outline was unaltered, the front roof-guttering had been removed, together with the welded strips which run from the rain gutter down each quarter of the body. The front and rear bumper-mounted flanges had been cut off, together with the front grille surround. The overall result probably does little for the car in terms of performance, but it certainly does give a much cleaner and more tidy appearance.

Jan told me that this had been done purely for amusement, though I find that hard to believe. Most good engineers like their products to be aesthetically right as well as mechanically perfect.

All windows including the front screen were made from perspex and the bonnet, boot and doors were of glass fibre. I was particularly impressed with the boot lid; no single skin shell this, but an exact replica of the standard article inside and out. This of course made it far more rigid than the usual lightweight rubbish on the market today. Similarly the doors had been very carefully boxed-in to increase their rigidity and at no time did I notice any difficulty in opening or shutting the doors, which is more than can be said of many saloons seen around the circuits these days.

Apart from an extra-large oil cooler which protruded through a hole in the grille, and an obvious lack of bumpers, the only other outward differences from a standard Mini were the wheels and flared wheel arches. The wheels were Tec-Del Minilite 5½ in. variety and they were fitted with the latest Dunlop R7 dry tyres. The neatly-flared wheel arches prevented the tyres protruding beyond the bodywork. Jan had built the arch flares himself using Tetrosyl and steel strips.

The suspension was of the dry variety and was lowered and fitted with Armstrong Adjustable shock absorbers. The rather excellent Rose-jointed Cooper anti-roll bar was fitted to the rear.

The wheels showed negative camber all round. This was obtained by modifications to the rear subframe and by lengthening the bottom suspension arm at the front. Jan has tried altering the front castor angle but found that with a limited-slip differential the resultant handling characteristics were most odd. Consequently no modifications were made in this direction. The engine was a normal 1275 S unit, bored 20 thous. to 1293 c.c. and fitted with dished pistons (Jan does not like the "Flat top" variety); The ordinary (not steel) flywheel was lightened and the clutch was of the competition diaphragm type. Needless to say the whole lot was balanced, and the cylinder head was gas-flowed. The valve gear had been lightened. Naturally enough the exhaust manifold was of Janspeed manufacture. I was most surprised to find that Jan was still using an "old fashioned" 45DCOE Weber carburettor, not the "latest" IDA variety. "Can't afford vun", he said.

Since a five-speed box is only an unnecessary luxury and a

possible source of trouble on a "Big Banger" "S", the more normal B.M.C. 4-speed, straight-cut close-ratio gears were used in conjunction with the 3.94 final drive ratio and a last year's limited-slip differential. Unlike most people, Jan has never experienced much trouble with these latter units. Could it be he incorporates a few extra mods of his own to improve reliability?

The drive-shafts were of the ordinary rubber jointed type, though a few days after the test the latest B.M.C. Hardy Spicer Universal jointed type were fitted.

In the "cockpit" one found a strong-looking roll-over bar running the whole width of the car. This again was of Janspeed manufacture.

An extremely comfortable cloth-covered bucket seat was fitted for the driver. The passengers' seats were no more than gestures, the front one merely consisting of very flimsy fibre-glass shell.

The "dash-board" had been panelled in matt black painted aluminium and into it were fitted oil pressure, water temp. and rev-counter dials. A speedo reading in K.P.H. nestled in the passenger side (heaven only knows why).

A 13 in. leather covered alloy steering wheel was fitted and the column raked down 1 in.

The gear lever was cranked back to fall more naturally to hand to compensate for the rather more reclining driver's seat.

No interior trim whatsoever was fitted.

Closer inspection inside showed evidence of the body-shells vintage and thinner construction. Several plates had been welded into the floor, obviously where badly rusted areas had been cut out, and in two places daylight showed through the floor where the cumulative effects of corrosion and abrasion from the driver's feet had caused tiny holes to appear.

However this was no fault of Janspeed and the impression throughout was of painstaking preparation and careful attention to detail. The whole car being otherwise absolutely immaculate, even to a good polish after it came off the trailer (it had one before it went on the trailer as well).

We were met at the circuit by the well known "character" Janspeed driver Geoff Mabbs. He was present to carry out the initial testing as this was in fact the first time that the car in this form had been driven.

Firstly Geoff went out to do a few laps and what a few laps it was. Never having driven the car he lapped after 10 laps ½ a second inside the class lap record, to see him drifting sideways in front of the pits was a sight for sore eyes. After getting inside the record things suddenly went silent and The Mabbs went missing. We were most worried and dashed out onto the circuit to find him. Fortunately car and driver were both safe but had experienced a most unusual mishap. The coil had suddenly failed completely, it was absolutely stone dead. Neither I nor Jan nor anyone else present had ever known this to happen so suddenly or completely before.

After towing the Mini back and fitting a new coil it was decided that since I had never seen the circuit before, Geoff should drive me in the Mini, round the circuit a few times. Consequently I donned my crash hat and endeavoured to get comfortable in my flimsy fibre glass shell called a seat. With fibre glass doors and no safety harness I was quick to look for something to hang on to, having noticed a mischievous glint in the eyes of Geoff and the Guvnor.

Suddenly all Hell was let loose as the engine was started, gears selected and the clutch let in. We departed from the paddock road onto the circuit like a demented she-cat. The noise was fantastic so I thought and the suspension at first seemed absolutely solid, but that was my last chance to take stock, we had arrived at Quarry! "Oh Gawd" I thought, "what have I let myself in for?" The answer was one of the most exhilarating and possibly most frightening experiences of my life.

No quiet conducted tour this but a flat out thrash, using the whole circuit and plenty of grass as well, Geoff sitting well back and really getting down to his work, every corner was taken sideways and smoke poured off the tyres.

After a couple of laps I had recovered some of my composure and began to enjoy myself, if this is possible in a bucking, kicking, screaming banshee, whilst hanging on for grim death to the roll-over bar to stop oneself disappearing smartly out through the side of the car via the fibre glass door. My seat had collapsed and I had to stand (ha ha) up. A whispered reply "Yea, great man" and if anything we went even faster. Till at last I cried "Enough, enough, I'm satisfied". My arms had begun to feel like lead and I felt unable to hang on any longer.

Thankful to still be in one piece I clambered from the Mini, a very much chastened character, to learn that the two of us had lapped in 1 minute 17·9 secs. Only 1½ secs slower than the class record. Some performance with two up, and small wonder I was scared. Many people get scared just watching Geoff drive. I rode with him and he did not hold anything in reserve. It was no small consolation when Jan said "well you're a braver man than I, I wouldn't ride with him".

Ten minutes later I don't mind admitting I was still shaking in my boots and trying to regain my composure, when the boys looked round and said it was my turn now.

On went the helmet again and I deposited myself into the driver's seat.

My first impression was how comfortable and firm it felt, especially with my safety harness fastened, support in all the necessary places was perfect and the cloth upholstery prevented any tendency to slide around or perspire too freely. The gear lever fell quite naturally to hand as did the steering wheel and all other controls (or should I say to feet) I must say that the 13 in. steering wheel did seem a little too small at first glance and in fact this proved to be the case in practice, I much prefer the standard sized wheel. Mind perhaps it's because I am no six-foot muscle man. Somewhat gingerly I started the engine and looked for the rev-counter, only to discover that I couldn't see it, it was completely hidden by the steering wheel. Geoff had made the same complaint since we are both about the same size this was not unexpected, and explained why the existing driving position suited me perfectly. As I started the engine I was once more aware of the tremendous din inside the car and a lot of vibration transmitted not from the engine but from the exhaust system, however, with no attempt made at fitting a silencer and a complete absence of any trim, this was hardly surprising. With the 649 camshaft there was virtually no tick-over speed and the revs needed to be kept up to prevent stalling.

I found, however, that I could pull easily away from rest at a little over 1000 r.p.m.

Immediately I used the loud pedal everything seemed to smooth out and I was no longer even aware of any suspension harshness.

The power really came in with a rush from about 3,800 r.p.m. and care had to be taken not to over-rev in the first two gears. Up through the box I went and suddenly I was into Quarry, a quick dab on the brakes, grab third and I was all crossed up at a diabolical angle to my direction of travel. Needless to say I at once clogged it and put on some correction. I was surprised to find virtually no understeer, even on full power in 3rd. Jan's suspension mods and the limited slip diff really work. This came as a most pleasant surprise, as did the discovery that one could "hang the tail out" at will, almost as with rear wheel drive, it was not necessary to back off to achieve this, merely one had to increase the lock. The nett result was that one could adopt almost any drift angle one wished and maintain or alter it whilst in a corner by adjusting the lock in conjunction with only the slightest alterations in throttle opening.

These characteristics served to give me the greatest confidence in one's ability to drive the vehicle, perhaps too much confidence because one felt that one could do almost anything, even perform the most ridiculous antics and still get out of trouble. One felt that there was almost no cornering limit. This of course was not true as I quickly proved by getting onto the grass, on the outside at Quarry next time round, this showed up yet another characteristic.

One hears tales of the Mini limited slip diff causing the car to veer as much as 6 ft. 0 in. to one side or the other of a straight line, depending whether or not one is accelerating or on the over-run. Frankly I found no trace of this, the whole vehicle being most stable at all times, and having exceptional directional stability. Only when I got two wheels on the grass did the diff exhibit any bad characteristic. I could not get off the grass. The more I accelerated the further on I got.

Finally, after about 100 yards I was forced to lift my right foot and immediately I literally flew back onto the track, nearly taking to the grass the other side. At about 100 m.p.h. this was not funny and I had to slow right down to get back on line for the bends at the back of the circuit away from the pits.

On reflection I could probably have avoided these alarums by taking my foot off more gently, but I decided against further experiment. There were no obvious signs of body roll from the driver's seat, though I had noticed some when watching Geoff doing his stint.

The gearbox was as sweet as a nut and cogs could be swapped as quickly as one could move one's hand without even backing off. The clutch was very positive but showed no signs of fierceness or slip.

Since I hardly used the brakes I can pass no other observations other than that they showed no vices whatsoever on the few brief occasions that they were used.

The engine was superlative being smooth, having bags of torque and almost unlimited top end power. Even at about 115 m.p.h. it was still pushing one back in one's seat, and in the gears one could almost snap one's head off with acceleration. I did not bother to take any figures as they would have been of little value depending as they did more on the driver's ability than anything else.

After a few laps I regretfully had to come in and pack up so that we could go home.

I can honestly say that at no time (other than the one incident at Quarry) did I feel that I was working hard, and credit must go to Jan in producing a car that is simplicity itself to drive, far easier than any other competition Motor I've tried. It was right first time. Some idea of just how easy the car was to drive can be gauged from the fact that I had only once ever been to Castle Combe before, as a spectator when I stood at the pits the whole time. Until that day I had no idea even where the circuit went, yet after only 3 or 4 laps I was circulating at about 1 minute 21 secs without really trying too hard.

The true potential of the car has since been demonstrated by Geoff Mabbs, the following weekend the car won its race outright in the wet at Combe and has gone on to win its class or race every time out excepting at the last Goodwood meeting when Geoff overdid things at the chicane and clobbered the kerbstones, causing the petrol tank to come loose in the boot and necessitating his retirement. He did no damage to the suspension etc. contrary to statements widely made in other motoring journals.

Truly a fabulous day's fun and most enlightening. My warm Mini Traveller seemed like a bath tub on the drive back from Salisbury, but it certainly got up to antics which I would never have believed it would have tolerated before. Mind you, I was Stirling Moss for a few hours.

This track testing is all go folks—Jan in the centre, Geoff on the right.

MINI-SPRINT GT

Stewart and Ardern transform the Mini bodyshell

WHEN the Mini was introduced most people thought a family saloon car could not be made any smaller if it was to seat four people in reasonable comfort, but someone always has to prove public opinion wrong and the first to do so, we believe, was Neville Trickett. His idea of lowering the car has been put into production by Stewart & Ardern Ltd., the Morris distributors, who skilfully cut the body in two places to reduce overall height by 4¼ in.

In fact the demonstration car is 5 in. lower than standard since the Hydrolastic suspension has been dropped, and it drew as much attention in the street as some exotic American cars we have been testing recently—not presumably because it is so startling, rather because it is obviously a Mini . . . but different. There is a definite trend toward "personalised" popular cars and this one might well catch on in limited quantity despite a large price label which reflects the amount of surgery done on the bodywork.

The basic Mini-Sprint conversion, which costs £335 onto the price of a Cooper or Cooper S, involves cutting the body above and below the waist-line, taking out a depth of 1½ in. at each level and increasing the rake of front and rear windows. The Mini-Sprint GT conversion costing £400 takes an extra 1¼ in. from above the waist, making a total of 4¼ in., removes all the welded seams which stand proud of the normal bodywork, replacing them by butt or lapped joints which present a neater and smoother appearance. Perspex is fitted in all side and rear windows (the back side-windows no longer hinge outwards) and of course the doors, wings, and boot-lid have all been cut to size. Included in the specification are an acrylic paint finish, in any colour, rectangular Cibie headlights, a small diameter leather-trimmed steering wheel and column rake adjustment.

This is obviously a very sizeable operation and the workmanship is extremely good; it is not possible to see where the cutting has been done. However, despite special optional front seats and rear seat trim, we did not feel that the general appointments of this car lived up to a £1,000-plus price tag and would prefer to spend some of the money on Radford treatment such as nice, well-fitting carpets, soundproofing, electrically-operated windows, an instrument console and so on to make the car feel more luxurious.

It looks like a Mini but . . . the bodywork has been lowered five inches by skilful surgery, and seams removed. Headroom is just adequate for a six-footer.

Wind noise has been slightly reduced by the elimination of seams, but the 1275S engine in the demonstration car was such a rough, noisy unit—as they all are—that it feels an exciting experience anyway to reach 100 m.p.h. which we estimated to be the maximum speed, an extra three or four miles an hour deriving from smaller frontal area and slightly reduced weight.

Lowered suspension adds to the overall effect but while improving the high-speed handling, it was not ideal for roadwork as the Mini-Sprint was really harsh on bumpy roads and the front wheels sometimes hit the wheel arches. The wheels, incidentally, were magnesium alloy with 4½J rims costing £70 for a set of five.

Individuality costs money and with a car like this it is hard to assess a reasonable price. Considering the work involved the cost is undoubtedly justified, especially since the bare bodyshells can be bought for £230 or £260 depending on the conversion and built up from there, and while we can think of better ways of spending up to £1,250 this is very much a matter of taste.—M. L. C.

IF in our road test of the standard 1275 Mini Cooper S we were disappointed because it wasn't an eccentric, burbling, rough-hewn hot shot then our Mr Hyde instincts were satisfied when we road tested a 1310 cc Cooper fully works prepared by BMC for rallying. This fantastic little beast was all we had expected. The engine flew into life with a right rorty exhaust note, the ride was like a billy cart over cobblestones, the exterior paintwork was done in vivid Castrol green and the clutch required Herculean effort to operate it.

With trialling the up and coming form of motor sport we thought a road test of this consistent top runner of the 1966 season appropriate. Our particular car was one of a team of three. Crewed by BMC Competitions Manager Evan Green and navigator John Keefe the Cooper rewarded BMC with third place in the NSW trials championship behind runner-up Bob Holden in another of the team cars. Our test car had also won its class (1300 to 1500 cc improved) in the Rothmans Southern Cross rally.

The right combination of reliability and speed is the most prized quality in a trials car. BMC has achieved this with its rally cars by hotting them to an extreme, then gradually de-tuning them until an acceptable compromise is reached. BMC has learnt that only the fastest finisher wins and the factory doesn't sacrifice reliability for sheer outright performance, although its cars are probably the quickest on the trials scene.

We try out BMC's proven rally winner

Take the first left-hander heading east: aircraft-type compass between navigator's legs keeps car pointed in the right direction. Note different seats.
Var-room. Rally S-type, all winking spot lamps and wire-mesh lamp covers over a jump-up and into full-flight.

Our car, EFT 685 got the full grind through the mill to arrive in the condition we road tested. The 1275 block was taken out to 1310 cc, twin 1½ in. SU carburettors replaced the stock 1¼ in. units and wild camshaft and porting and polishing completed the picture. The result was a cammy, cranky little car which still showed our Renault entry a nifty pair of heels in all the latter-season rallies last year and that gorgeous blurp-ety blurp-ety note of the cars idling in controls was perfect demoralisation tactics for the opposition.

But it was all very impractical too, and during the last rally year the cars were made more docile by reducing the compression from near 11.5 to 1 to 9.5 to 1; the wild cam gave way to a milder, less fussy shaft with a more subdued note. In this state we picked the car up from the factory just after it had run in the Victorian Alpine rally and been used for route charting the 1967 Mini Monte Rally. The wiring had been replaced after it burnt out on the Alpine but apart

A RALLY BOMB WITH A SHORT, SHORT FUSE SOOPER COOPERFORMANCE

SPECIFICATIONS "RALLY" COOPER S

MAKE .... Morris   **MODEL** .... Cooper S (rally car)
BODY TYPE .... Sedan   **COLOR** .... Castrol green
MILEAGE: Start 5667   Finish 5920
FUEL CONSUMPTION:
Overall ...... 22.2 mpg (26 mpg).   Trialling ...... 15-20 mpg (approx)
SPEEDOMETER ERROR: Indicated   30 mph   50 mph   60 mph
                    Actual   30 mph   50 mph   60 mph
PERFORMANCE
       (Standard 1275 cc Mini Cooper S in brackets)
     (All performance figures done without clutch — refer to text)
Top gear mph per 1000 rpm .... 12.9 mph approx
Engine rpm at max speed .... 7000 rpm at 90 mph
Lbs (laden) per gross bhp (power to weight) 14
MAXIMUM SPEEDS:
Fastest run .... 90 mph   (96 mph)
In gears: 1st 33 (37), 2nd 50 (59), 3rd 70 (84), 4th 90 (96)
ACCELERATION:
0-50 mph .... 8.0 secs (8.2 secs)
Fourth gear:
20-40 mph .... 8.5 secs (7.3 secs)
30-50 mph .... 9.2 secs (7.4 secs)
40-60 mph .... 8.6 secs (8.6 secs)
50-70 mph .... 9.4 secs (NA)
STANDING QUARTER MILE:
Fastest run 18.4 secs (17.6 secs).   Average of all runs 19.2 secs (17.9 secs)
ENGINE:
Cylinders .... four in line
Cubic capacity .... 1310 cc
Compression ratio .... 9.5 to 1
Valves .... overhead pushrod
Carburettors .... twin 1½ in. SU
Fuel pump .... high pressure SU
Power at rpm .... 100 to 105 bhp
TRANSMISSION:
Overall ratio .... final drive 4.133 to 1
CHASSIS and RUNNING GEAR:
Suspension, front .... hydrolastic displacers high pressure
Suspension, rear .... hydrolastic displacers high pressure

from that the car had not been touched. This made for a very poor state of tune as the accompanying figures show, but an out-of-tune 1310 cc Cooper will still hold its own with most machinery around town.

The interior fittings are completely functional. The normal brightwork hardware was carefully sprayed anti-glare matt black: speedo trim ring, switch panel, tacho and Halda Speedpilot housing all got the treatment. The steering wheel spokes were bound with tape. About the only shiny area not blackened was the windscreen and the fire extinguisher.

Below the normal line of switchgear on the parcel shelf was a row of five toggles, four to control the auxiliary lamps and one the washers. The lamp switches are wired so that high beam cuts all. Presumably the lamps were wired separately so that if one shorted out after an accident it could be turned off separately. Otherwise different combinations of fog and driving lamps to suit the season may have inspired them to use individual switches. Under the Halda Speedpilot were three switches and a jack. One switch we ascertained lit the Speedpilot face. The others? We never found out. There were also two switches in the console odds and ends box which seemed equally without purpose. The car had no reversing lamp so one switch would have been used on that and for sure they all had a purpose.

Two plus two (tyres). Spare SP44 shod wheels are bolted in place of rear seats with quick-release capstannuts. Not so good for more passengers.

All a Mini's troubles are in front. BMC rally Cooper sports sump pan, high mounted number plate to back up good lights and tyres. Front is very robust.

A Butler flexible navigator light was fixed to the roof so as to be above the navigator's head and the 7000 rpm Smiths tachometer was fixed in the parcel tray directly in front of the driver. Sponge rubber applied to the door corresponded to Mr Green's knee and shin positions and presumably was necessary to give some protection for rough going. The equipment was completed with two spare tyres bolted through their rims to the rear floor with large capstan nuts and a boot full of useful oddities, a Japanese fold away shovel, a tin of Hydrolastic fluid, a box with fan belt, spare headlamp, can of WD40 and sundry odds and ends. The most interesting discovery was a VW jack welded to suit the Mini's jacking point. There was a Mini jack as well. A tin of Castrolite and a large piece of sponge rubber to stop rattles filled what there was left of the boot which had to accept another five gallon tank as on all Cooper S types. The Hydrolastic suspension lines were led through the cabin to protect them against stone damage. Is that an admission of something? Under the bonnet there is nothing to see. The engine looks disarmingly like any other Mini Cooper S.

Out front four driving lamps (one broken on the test car) sit on a flimsy piece of angle iron which was found vibrated badly. The number plate locates itself on the bonnet lid (highly fashionable in the right circles) and an octopus strap between over-riders and number plate also helped to put

the Cooper among the in-group, in rally trends, as well as serving as an emergency bonnet strap. Dunlop SP44 Radial ply winter-tread tyres grace the 10 inch wheels and the car is supported on high pressure Hydrolastic units with the rear valves poking through inspection holes cut in the boot floor, for emergency re-pressurising.

Rally drivers need to be a cross between Mario Andretti and farmer Brown: fast but not ashamed of the occasional bit of gardening. The Rally Cooper is immensely suited because of the mild cam the greatest torque comes in between 4000 and 6000 rpm. The car uses the Morris 1100 4.133 to 1 differential which drastically undergears it — by intent. Combined with the torque band high up there is always dynamic power between 50 and 90 mph. Even in top gear the tiny wheels will spin furiously on loose gravel. Despite this, the car is still tractable at low speeds. It will chug off happily from right down at 15-20 mph in top as soon as the accelerator is prodded. By rights such a hot piece should ping and shudder — it doesn't, just gets up and goes, all the way to 90 mph.

Because of the undergearing the tachometer read a maximum 7000 rpm when the speedo showed 90 mph. We took this as top speed, abandoning our normal full windout runs because the car was still accelerating violently but at maximum rpm at 90. The clutch slave cylinder seals gave out early in the performance figures and we had

very little clutch left. By pumping furiously then grabbing first gear we were able to run the few performance figures listed. All gear changing was clutchless and considering the engine was tired, a standing quarter mile time of 18.4 seconds rated quite fast. Well set up we feel it would run comfortably under 17 seconds and that's getting down near Jaguar and Mustang times. Top gear figures show that times went down as speed increased. On the whole, times were slower than the 1275 cc Mini Cooper tested in February 1966 WHEELS but this was due to the faulty clutch and poor tune.

On dirt the Cooper is very, very quick. Gear changes come as a matter of thought and clutchless changes and left foot braking is natural. With the SP44s the handling is the closest to neutral a Mini will ever come. Only on wet bitumen is there excessive front end plough around corners. In this situation it is easy to pour on too much power so the wheels spin and increase the understeer.

The whole effect is very much rallye! The car is noisy with lots of the right vibration noises, the diff howls and sounds like a police siren, the gearstick rattles enthusiastically if not held and the four lights and the Castrol stickers bring second looks everywhere. You feel you're not doing the right thing until with clenched teeth and grim determination you grab three at 75 with the passenger yelling "left point two five, caution, through ford and into control."   #

The garbage compartment. Don't laugh, it's all necessary. Gear includes spare lights, fan belts, jacks and hydrolastic fluid as well as wire, rope and others.

Well-adjusted torque band and special rear end gearing keeps dirt spurting from those front wheels — even flat out in top. Handling the car is a breeze.

ONE way and another we thought it had been a bit of a long time since we took a careful look at the quick Minis—as they come out of BMC's production lines, that is. The quick ones for the purposes of this particular demonstration are the 998 Cooper and the 1275 Cooper "S", the first being a sort of halfway house between the three basic Minis and the second a kind of instant racer. In both cases—and this applies equally, come to that, to the 850 as well, although that needn't bother us here—they can be turned from quite quick, relative to their capacity, to very quick without too much in the way of radical modification: if you carry the thing to its conclusion they can both go ruddy fast to put it mildly.

In standard form, however, they are both sporting little cars, with good roadholding and admirable performance, although when ultimate top speed and acceleration are not of vital importance we would be inclined to settle for the Cooper 998 as a road car because although it doesn't do quite so much, it does what it does do in a pleasanter fashion.

The specification of the pair of 'em is similar in all respects except that of engine, transmission and brakes: the Cooper "S" is also a shade heavier, but the extra urge from the bigger engine more than takes care of that. Both are undisguisedly Minis, and at a quick glance only the badges and chrome lettering distinguish either of them from a bog-standard 850. A few distinctive touches in the available colour schemes, plus the aforementioned badges and so on, make sure the neighbours know you're a step ahead of them in the Jonesmanship race, but otherwise they are Minis. You get the same bodyshell, trim, badly-placed and limited instrumentation, dreadful seats and still poor heating and demisting arrangements on both of them, although the Cooper "S" we had for test was fitted with a pair of optional reclining front-seats which were not perfect by a long way, but which certainly represented a mighty improvement over the standard perches.

The engine of the Cooper has a bore and stroke of 64·6 mm. × 76·2 mm.; with a 9 to 1 compression ratio and two HS2 1¼ in. S.U.s it develops 55 b.h.p. net at five-eight and maximum torque, 57 lb./ft., is achieved at 3,000 revs. The four-speed gearbox is operated by a remote control change and has no synchromesh on bottom gear. The ratios, with the standard 3·77 final drive, are not exactly close-ratio but form a useful compromise: frankly, with the standard unmodified engine close-ratio gears would probably be a thought embarrassing with so little torque—although the torque available is, admittedly, a considerable increase over the 850. Standard ratios, in fact, are 3·77 top, 5·11 third, 7·21 second and 12·05 first: this top gear gives 14·7 m.p.h. per 1,000 r.p.m. so that at five-eight, maximum power, the car is travelling at 86 m.p.h. or so. All Mini gearboxes now have baulk-ring synchromesh and needle-roller bearings, and the only variation is an optional 3·44 final drive which gives 16·05 m.p.h. per 1,000 r.p.m. in top and which gives overall ratios of 3·44 top, 4·67 third, 6·59 second and 11·03 bottom.

Suspension of both cars is, in common with all Minis except the van-bodied models, Hydrolastic and independent front and rear, with wishbones at the front and trailing arms behind. The brakes on the 998 Cooper are disc on the front, seven inches in diameter, with seven-inch drums at the rear.

The "normal" Cooper, as we said earlier on, is probably the nicest to drive, albeit obviously far from being the quickest. Not exactly a powerful monster, 55 horse power is nevertheless enough to make it a very lively little car which can be tremendous fun—perhaps "peppy" is a good word. For a variety of reasons which we won't go into here, we had the

A COOP

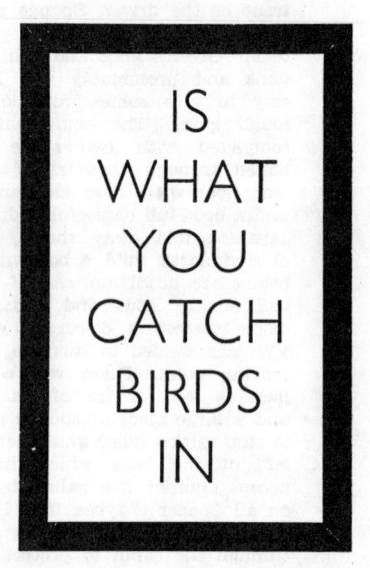

IS
WHAT
YOU
CATCH
BIRDS
IN

AND IN COOPERS YOU CAN CATCH ALMOST ANYTHING

test car on two separate occasions, once in ruddy cold weather and the second time in warmer, wetter climatic conditions, and under both circumstances it was always a perfect starter. The temperature gauge needle ran up to the "normal" position in under a mile and then stayed there, and the hardest driving and the thickest traffic didn't get it any further up the scale. The engine, which by modern standards has a relatively long stroke, and whose crankshaft runs in only three main bearings, is pretty smooth and sweet and never objected to a free use of high revs—in fact to get the best out of the Cooper you need to get the little old crankshaft fairly whizzing round, and while our total test mileage probably went well over the 2,000 mile mark, it never gave any trouble, nothing vibrated loose except for a piece of door trim, and oil consumption, at around 550 miles per pint, was entirely reasonable, we thought, when you consider how the poor little thing was thrashed about the place.

We didn't much care for the gearchange on this particular Cooper, which was stiffer than we felt was reasonable—so stiff, in fact, that enough muscle was necessary to move the thing about as to make it altogether too easy to overcome the reverse gear spring and make nasty noises from under the floor. The throttle linkage, too, was jerky—you hardly ever, in our experience, find a standard Mini with a smooth throttle operation—and the two things combined to make the car a shade awkward in traffic. The gearbox, though, is very much O.K. in spite of the anachronistic unsynchronised bottom gear which BMC is still—amazingly and unforgivably—foisting off on the market. Its ratios may not, as we said before, be the greatest thing in close ratios, but they provide a very reasonable compromise and she'll do 65 in third and 46 in second, so that they aren't as bad as they might be. The diaphragm-spring clutch takes a bit of getting used to unless you have recently been doing some Mini motoring, and is a thought "in-out" to catch the unwary the first time you move off. So far as incompetent us are concerned, it meant an undignified stall at the BMC depot under the eyes of altogether too large a grinning audience. The Hydrolastic suspension has, to be honest, never really seemed to be worth the fuss. True, it reduces the bouncing and shaking you used to get on "dry" Minis on poor surfaces, but against this there seems to be a lot more pitch, and the difference in attitude—nose up or nose down—under acceleration and braking is great enough to be noticeable to by-standers. To passengers, it is even more pronounced, and on the test car was emphasised by the jerky throttle. You get SP41 tyres on the Cooper, and no-one can deny that these plus the suspension set-up and in-built f.w.d. characteristics do provide extraordinary roadholding and cornering performance. The steering, of course, is delightfully quick and precise, and although the extra power over the 850 Mini emphasises the unusual aspects of the handling it also makes it easier to control when you're thrashing it about the country lanes the way the Cooper was meant to be thrashed. Understeer with the power on is increased, but so is the transition from understeer to oversteer when you lift off more marked, and this means that a wider range of tricks can be played. To get into trouble with a Cooper would be very difficult, and anyone who could do it would be a pretty good sort of incompetent idiot. The SP41s hang on very well under wet or dry conditions, but once you've got the hang of the thing you can take advantage of a wet surface to slide the car under perfect control—the sort of behaviour which isn't to be recommended on crowded public roads, but if you can find somewhere safely private it is the most enormous fun.

The biggest disadvantage on the 998 Cooper raises its ugly head when you come to stopping the beast. The seven-inch disc brakes are, admittedly, better than those fitted to early Coopers—they needed to be—but they still provide remarkably little stopping for remarkably high pressure. So long as you don't overdo it, however, they will stop the car. But if you use 'em too much too often they fade right away, and fast driving where other traffic is has to be an exercise in thinking ahead if your pride and joy isn't going to end up with a short wheelbase.

Any other snags? Well, as we've said already, the driving position is really nasty, with a badly-designed seat, a poorly-placed steering wheel and minor controls positioned in such a way that, with the seat positioned to accommodate comfortably a driver of average size and tastes—our Ed. for instance—you can barely reach the unidentified switches over there on the middle of the dash. In the same way, everyone who drove the car while we had it found that when holding the wheel in the normal position you couldn't see the oil pressure gauge, nor half the speedometer. It is also noisy inside, and lord, how we hate those sliding windows! But like the man said, "Vot you expect for the money—my life?".

Cooper 'S'

The Cooper "S" is like the Cooper only more so. In other words, it is still just as much a Mini, but it is faster, livelier, thirstier and noisier. In standard form it is just as uncomfortable, too, but remember "our" car had optional reclining front seats which helped. The engine in this case is completely different—it has offset bore centres, the block is taller and the bore and stroke, at 70·6 mm. × 81·3 mm., give a total capacity of 1275 c.c. Compression ratio is 9·75 to 1 and there are two S.U.s again. There are other internal differences, including Duplex timing gear, crankshaft bearing sizes, larger valves and so on, and the ultimate difference is in the power—76 b.h.p. net at six thousand, and 79 lb./ft. maximum torque at three thousand. A wide variety of optional gear ratios is obtainable, although as standard the car comes with the 3·7 final drive and so on, similar to the Cooper.

This is in every way much more of a sporting carriage, and is a good deal quicker in every respect. A mean maximum of 99 miles an hour against 88 for the Cooper; acceleration from 0–60 in 11·6 seconds and 0–80 in 24·5, against 14·8 and 34 seconds for the Cooper, with over eighty available in third and close on sixty in second—it is really quick by any standards, is this Cooper "S". But you sit yourself in state in the self-same poorly-instrumented and cheaply-furnished interior, and even with those optional seats—more comfortable, admittedly—the driving position was very little better. Back-ache was much reduced, but the steering wheel position is still miles away from most people's taste and you still can't reach the still-unidentified switches in the middle panel without leaning forward. And the engine is among the noisiest and roughest in current production. There's no rev-counter, but the speedometer dial is marked for the maximum speeds in the gears, including top. And if you get within five miles an hour of any of them the motor gets

Cooper

as rough as a bag o' nails and the tendency is either to lift off or change up, according to which gear you're in and the situation at the time. Probably this is good (and very subtle on the part of BMC) for engine life, but hardly has the same effect on maximum performance, and this is undeniably a performance car. Although it is, admittedly, heavier (though not by much) than the Cooper there's a helluva lot more urge to compensate for it, and the thing steps off and along in sharp old fashion. The track is wider because of spacers and wide-rim wheels, so that handling is outstandingly outstanding instead of just outstanding. As with all Minis the more power you've got the more understeer you get with the power on, and as most of the Cooper "S" weight penalty is in the engine, which means at the front, this would conceivably give problems in the transition from under- to oversteer when the urge is taken away if it wasn't for the extra track width. As it is, the whole thing works perfectly.

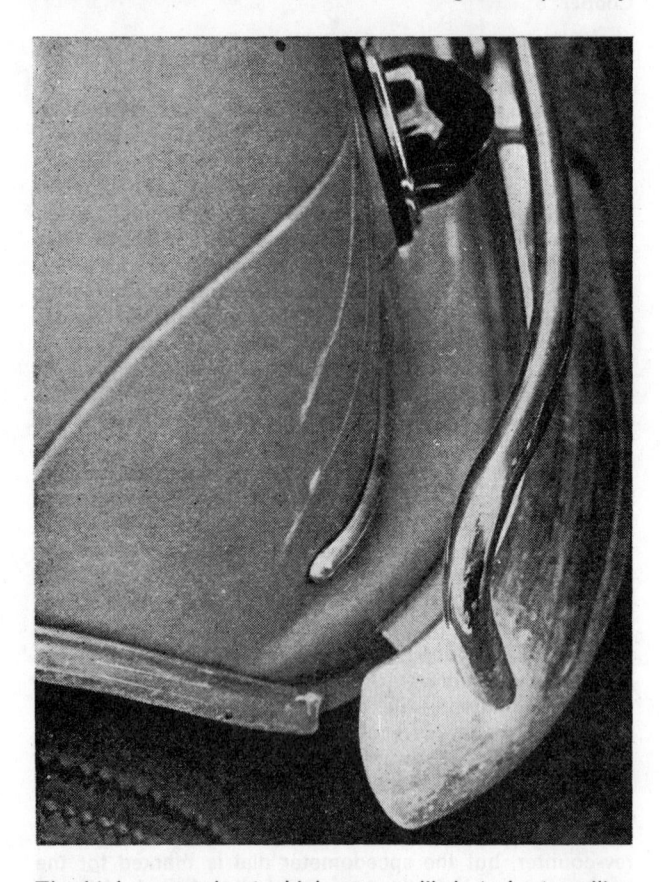

The higher speeds at which you are likely to be travelling give rise, on the Hydrolastics, to a good deal of body movement, and if this isn't exactly big-car stuff at least the movement is well-damped: false teeth will stay in, and sun-glasses on, even if well-built young women provide a spectacle which can be dangerously distracting (we are told).

You can put the Cooper "S" exactly where you want it under almost any conditions or circumstances; this is not only an advantage but is also highly enjoyable, and the degree of control which even quite average drivers (you don't *have* to be Paddy Hopkirk—it just helps) can exercise is obviously a major safety factor in what must be a very safe car.

Safe? Certainly—it even stops. Bigger disc brakes and modified calipers provide stopping power which is well up to 100 m.p.h. performance—and nothing to laff at at all. On the go-side, you could do with an extra ten miles an hour in second for the best acceleration, but the increase in

torque helps a good deal and there is no ghastly hush immediately on changing from maximum revs in second to nothing much in third.

Going, stopping and handling aren't necessarily all you care about in a high-performance car like this one, which is some way from being cheap. Uncomfortable, poorly-shaped seats, a very high noise level, an abominable driving position, poor instrumentation, a ridiculous horn, plus the lack of such basic items (by modern standards) on a fast car as a headlamp flasher (especially when presumably the one from, say, the Austin 1800 could so easily be made available) and self-parking, two-speed wipers seem curious omissions on a car which, in price or performance, compares with the Sprite, Fiat 850 coupe, Abarth OT 850, Cortina GT, Sunbeam Alpine, or the Imp Sport, all of which are relatively refined cars when it comes to equipment—some very refined indeed.

In spite of all this, however, the Cooper "S" is definitely a car of character. You love it, hate it, or generally feel strongly about it—but no-one is indifferent. And that, now that cars are beginning to look, feel and perform in much the same way, is a plus-point in itself for our money. Too often (for our taste) you almost have to get out and look before you can remember what car you're driving. But this doesn't happen with the Cooper "S". If it's performance you want— and if you are prepared to put up with some discomfort and some inconvenience—then the Cooper "S" must be a Good Investment. If you don't want quite so much performance—or if you can't really afford it, or the insurance company won't let you, then the 998 Cooper takes a bit of beating, too. One thing is sure: attempting to follow a well-driven example of either along a twisty road is not something to be taken on lightly!

Baby GT powered by Mini-Cooper

THE Unipower is rather different from most of the other specials which use BMC Mini components in that any part which is not the right shape for this concept has been replaced rather than modified. We thus have a standard Mini-Cooper engine in the rear together with various of the front suspension components used in a different way to give wide-angle wishbones a la formula cars.

The concept is really a road-going two-seater Formula 3 car with as much use of standard BMC components as will keep the kit price down to a reasonable level—£950 for the car as we tested it. It is not quite as impractical as this might suggest; the engine is a bit noisy and there is no space inside the car for anything other than people, but there is quite a reasonable insulated luggage compartment just behind the engine. The engine is still mounted transversely and is in front of the rear "axle" line; the gear linkage comes down the right side-member and the gate is a standard Mini one turned through 180°.

A space frame chassis is used with integral roll hoops and glass fibre panels are bonded to the structure, which thus receives some stiffness from them. The rear half of the body hinges backwards for access to the engine and luggage; the front boot houses the petrol tank and spare wheel and gives access to the pedal box with its three fluid reservoirs—twin circuit braking is used.

The whole design is well finished and thoroughly roadworthy once you remember that in traffic you are frequently too low to be seen by bigger brethren; eye level is about the same as the door handle of a Mini. The racer's reclining seat position is easy to get accustomed to and seems entirely natural on the open road; you can lean forward in towns to look round from a side turning if you aren't wearing belts.

With a weight saving of 1.4 cwt. over a standard Mini-Cooper plus a better shape and lower frontal area it is not surprising that the performance is rather better too, despite this car having the 3.44 axle compared with the standard Cooper 3.765. Top speed is over 100 m.p.h. and fuel consumption is remarkably good with over 50 m.p.g. at 60 m.p.h.

The roadholding is about as good as one might expect for such a layout designed by idealists and you have to be trying quite hard to reach the limit characteristics on the open road; unfortunately the ride has suffered accordingly, being very firm—uncomfortably so around town although it gets better at higher speeds—and allowing far too much pitching which prevents one using the head-rest on any but the better roads. High-speed running is not truly stable either, since the car wanders in the slightest of side winds.

The main question is—does the Unipower do what it sets out to do? Generally, yes—it gives interesting driving and rewards you for the effort you put into it, much more on the open roads, of course, than in town where it is hardly practical for shopping, particularly because it is a difficult car to get in to. At £950 it is fair value for the enthusiast who still likes to combine acceptable road transport with a car that will not be outclassed at the weekend sprints, particularly with the 1275S engine.

Engine and transmission

The choke is behind your left ear on the engine bulkhead and is quite easy to use. Being just a standard Mini-Cooper unit with only a few differences for installation's sake—exhaust system and pancake air filters—it started easily; with long battery leads the engine only fires when you release the starter. A front mounted radiator, with a separate heater circuit as well, requires a lot of water in circuit and it takes a long time for the unit to reach running temperature. It would probably be quicker if one turned off the heater circuit with the temperature control which is alongside the choke—behind the passenger's right ear. An electric fan with a tell-tale light provides cooling in traffic.

A comparison of the figures of the Unipower and the standard Mini-Cooper provides an interesting compound study of the inter-effects of drag, gearing weight and frontal area but, despite the gearing penalty—a 9% loss of starting torque plus the impossibility of getting wheelspin with the rear engine—the benefits of a 9% weight saving under test conditions are apparent even in the 0-30 m.p.h. time with 4.1 sec for the Unipower against 4.5 for the Mini-Cooper. Beyond that the shape and frontal area have increasing effect; 60 m.p.h. is reached in 12.6 sec against 14.8 and 80 m.p.h. in 23.9 sec against 33.6. The top gear accelerations record a similar pattern except that the Mini-Cooper is slightly quicker from 20-40 m.p.h.

Continued on the next page

Baby GT powered by Mini-Cooper
continued

". . . eye level is about the same as the door handle of a Mini . . ."

We originally tried the car with the optional 2.9:1 final drive ratio, but felt that this was higher than people would want with this engine. The "long legged" feeling in top gear was not really adequate compensation for the rather sluggish getaway. Rather to our surprise the top speed was limited by valve bounce and a lap of the MIRA circuit at 100.8 m.p.h. included easing the throttle on a slight downhill section to keep under 6,300 r.p.m. (indicated); we had thought that the Cooper engine would be unable to pull the 2.9 axle to peak power but we now suspect that the Unipower is more likely to approach 110 m.p.h. on the 2.9 axle since this represents revolutions rather nearer to maximum power than our over-the-top revs on the 3.44.

Our fuel consumption of 31.5 m.p.g. is exactly the same as the Mini-Cooper's but in the case of the Unipower this was taken over a shorter mileage in which testing is a greater proportion; this includes a lot of driving on a closed track at speeds higher than the legal limit. In view of the touring figure of 47.1 m.p.g. which corresponds to 65 m.p.h. cruising, it should be quite possible to return over 40 m.p.g. with hard road driving.

The ratios are as for the standard Cooper which gives useful maxima in the gears, second being particularly handy for overtaking. Once we got used to the gate, we found the gearchange very pleasant with short precise movements around well defined slots. The lever is on the right, sticking out of the door sill and ideally placed to disappear up your trouser leg as you wriggle in, but also well placed for the right hand to drop on to it; elbow room is limited and shorter people, with the seat forward, will find the lever a bit behind the ideal when in its rearmost slots. These are in 1st and 3rd gears; you push straight forward from 1 to 2 and from 3 to 4. Despite the unsynchronized first we found it quite easy to double de-clutch sufficiently accurately on the move. We would have liked somewhere to put the clutch foot since the bulkhead is too far behind the pedal to provide a brace. A fair amount of final drive whine reaches the interior.

Roadability

When we first tried the Unipower for a few days we thought that it was almost impossible to lose adhesion on dry roads, but as we later became accustomed to it and tried harder we found the limit in easily controllable oversteer at quite high cornering forces which one might expect from a 40/60 front/rear weight distribution. Double wishbone suspension is used all round and it is set up to make the car behave very well on dry roads; the steering allows some kickback and is nice and direct if a little woolly and insensitive in the straight ahead position. This makes it difficult to feel loss of adhesion except with the seat of pants, although it starts to slide sufficiently gradually for this to be quite safe.

The suspension is adjusted to give negative camber at both ends and a fair amount of toe-in at the rear (an understeering stability factor). Once the car gets into the corner and roll is generated, the outside tyre tread is placed flat on the ground and the grip is good, but if the surface is very slippery (like the first shower on greasy London streets) there is not enough initial grip to get sufficient roll for the treads to flatten on to the road and the result is most unstable; you are not sure which end is going to go first and how far. Unfortunately this tends to colour one's whole impression of wet road driving unfairly since on purely wet roads the Cinturatos grip very well and the behaviour is very much as on dry roads—controllable oversteer.

Sidewinds have a surprising amount of effect on the car at speed

The semi-reclining seats require some contortion to get into, particularly by the driver, but are comfortable once in although the headrest is too springy.

Simple facia has been designed to be more functional than stylistic and requires either considerable familiarization or Dymo tape.

Some luggage can go in the compartment behind the transverse engine. Double wishbone and coil spring suspension is used all round.

and even with little wind there is still a slight tendency to wander; this is magnified when the luggage compartment is loaded and the tyre pressures should probably be adjusted for this condition.

At speed on smooth surfaces the ride feels quite pleasant and you can relax back on to the full length reclined seat; undulations are taken well and even sharp bumps seem quite remote, but at lower speeds in town radial thump comes through strongly and the car pitches. This is a very short movement but feels rather more if your head is resting on the backrest since each pitch kicks your head forward again—it is best to keep it forward which is no strain. There was not enough shoulder support either for good location with fast cornering. Visibility forward is good and not bad through the rear mirror which has little more than a slot to aim through, but the rear quarters are unpleasant blind spots in heavy traffic.

Interior

We understand that a more efficient silencing system will replace the present one which should reduce the noise to an acceptable level; currently it is buzzy under hard acceleration and at high speed, but really quite reasonable at a steady 70 m.p.h. since the engine is hardly working at all—witness the steady speed consumption of 48½ m.p.g. There was little wind noise and we found it fairly draughtfree with the side windows open.

There is one fan for both the radiator cooling and the heating system, the idea being that the only time you are going to need the heater boosting fan is when you are in traffic, and you will probably need the cooling fan as well; this is sound reasoning, but it rather reduces the flow through the heating system, which doesn't get very hot nor even very cold whatever you do to the water valve. Since our test further modifications have improved the flow considerably, and control should be rather greater. There is still a minor snag, that the low intake can let in fumes from the car in front; a flap controls heat to leg level leaving the rest to go to the screen.

The personal preference of one man is evident in the siting of the controls and this has been done without thoughts of styling—very laudable. The speedometer has been placed on the left for the sole edification of the passenger although not out of the driver's way. Rev counter and the oil pressure/water temperature gauge can be seen through the top of the three-spoked wheel and the fuel gauge (with an undamped needle) is on the right. The right hand stalk flashes the headlights (up) and dips (down) and the left hand one is for the non-self-cancelling indicators and the horn.

We were initially a little confused by the switches; three in the centre for which "up" is "on" and two on the right; the left of these is a two-position one for the lights ("on" is "down") and the right hand one is for the panel light with "on" being "up", but since this one is hardly ever used there is no great confusion on the right. Nor is there in the centre, once you realize that each switch is under something which has a direct relationship. From left to right, the fan switch is under its warning light, the interior light switch is under the light itself and the wiper switch is under the washer button; all very simple really. **M**

Performance

Performance tests carried out by *Motor's* staff at the Motor Industry Research Association proving ground Lindley.

Conditions

Weather: Dry light winds 8-14 m.p.h.
Temperature: 46°-52°F. Barometer: 29.1 in. Hg.
Surface: Dry concrete and tarmacadam.
Fuel: 4-star 98-octane minimum.

Maximum speeds

	m.p.h.
Mean banked lap	100.8
Fastest ¼-mile	102.8
3rd gear } at 6,000 r.p.m.	70.5
2nd gear } at 6,000 r.p.m.	50.0
1st gear } at 6,000 r.p.m.	30.0
Maximile speed:	
Mean	100.5
Best	102.2

Acceleration times

m.p.h.	sec.
0-30	4.1
0-40	6.3
0-50	8.7
0-60	12.6
0-70	16.8
0-80	23.9
0-90	33.9
Standing ¼-mile	19.1

m.p.h.	Top sec.	3rd sec.
10-30	—	—
20-40	10.4	6.5
30-50	9.8	6.5
40-60	9.9	6.8
50-70	10.6	7.4
60-80	12.8	—
70-90	17.8	—

Fuel consumption

Touring (consumption midway between 30 m.p.h. and maximum less 5% allowance for acceleration) . . . 47.1 m.p.g.
Overall . . . 31.5 m.p.g.

Steering

Turning circle between kerbs:	ft.
Left	40.3
Right	37.7
Turns of wheel lock to lock	2.1

Parkability (gap needed to clear a 6 ft. wide obstruction parked in front) . . . 5 ft. 2 in.

Weight

Kerb weight (unladen with 50 miles-worth of fuel)	11.2 cwt.
Weight laden as tested	15.0 cwt.
Front/rear weight distribution	39½/60½

Price (car prices refer to component form)

Standard road car with 998 c.c. Cooper engine and 3.44 final drive	£950
Road GT with 1275 c.c. Cooper S engine, S brakes and reset suspension	£1,145
Car as tested with Cosmic wheels, heater, sun visors, jack and one seat belt	£1,008 10s.

Extras

Heater/demister unit	£17 10s.
Car jack	£2 15s.
Oil cooler and connections	£12 10s.
Adjustable shock absorbers	£30 0s.
Sun visors	£3 0s.
Seat belts (inc. fitting)	£10 10s.
Cosmic alloy wheels (when supplied as original equipment)	£30 0s.
High final drive (2.9) and speedo	£39 10s.

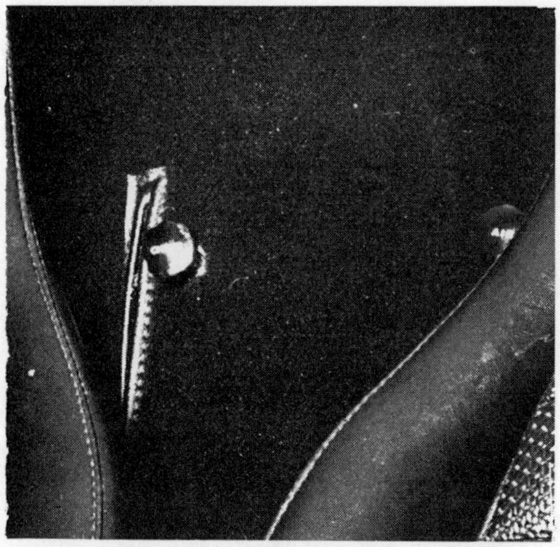

"The choke is behind your left ear . . . heater water valve is behind the passenger's right ear . . ."

Specification

Engine

Cylinders	4
Bore and stroke	64.58 mm. x 76.2 mm.
Cubic capacity	998 c.c.
Valves	pushrod o.h.v.
Compression ratio	9.0:1
Carburetters	Twin HS2 SU
Fuel Pump	SU electric
Oil filter	Full flow
Max. power (net)	55 b.h.p. at 5,800 r.p.m.
Max. torque (net)	57 lb. ft. at 3,000 r.p.m.

Transmission

Clutch	7⅛ in. s.d.p.
Top gear (s/m)	1.0
Third gear (s/m)	1.36
Second gear (s/m)	1.92
First gear	3.20
Reverse	3.20
Final drive	Helical spur 3.44
M.p.h. at 1,000 r.p.m. in:	
Top gear	16.0
Third gear	11.8
Second gear	8.3
First gear	5.0

Chassis

Space frame with bonded glass-fibre body

Brakes

Type	Twin circuit disc front, drum rear
Dimensions	7½ in. discs, 7 in. drums

Suspension and steering

Front	Independent: double wishbones and coil springs
Rear	Independent: double wishbones and coil springs
Shock absorbers: Front } Rear }	Telescopic
Steering gear	Rack and pinion
Tyres	Pirelli Cinturato 145—10
Rim size	4½J—10

Dimensions

Overall length	13 ft. 8 ins.
Width	4 ft. 9 ins.
Height	3 ft. 4.6 ins.
Wheelbase	7 ft. 0 ins.
Front track	4 ft. 0.8 ins.
Rear track	4 ft. 1.6 ins.

Figures taken on wet track and not comparable with Autocar Road Test figures. 0-60 mph times could not be obtained on cars marked with asterisk since our electric fifth wheel speedometer could not be fitted

OSELLI
850 c.c. Mini to Mini 7 Club formula.
¼-mile 18.1sec: 0-60 mph 12.3sec

MARCOS
Mini Marcos GT; 1,323 c.c with Janspeed head and camshaft.
¼-mile 17.5sec: 0-60 mph*

OSELLI
Stage 1 Morris Mini-Cooper S; 1,275 c.c.
¼-mile 18.9sec: 0-60 mph 12.8sec

DOWNTON
Stage 2 Morris Mini-Cooper 998 c.c.
¼-mile 20.8sec: 0-60 mph 16.9sec

Drag day for hot Minis

By
**Martin Lewis and
Michael Scarlett**

AUTOCAR 3 April 1969

UNIPOWER
1,275 c.c. standard
Cooper S engine.
¼-mile 18.3sec:
0-60*

BMC-DOWNTON
Stage 1 Austin Mini
Mk.II 998 c.c.
¼-mile 22.3sec:
0-60 mph 21.7sec

DOWNTON
Touring conversion
Austin Mini-Cooper S
1,275 c.c.
¼-mile 18.2sec:
0-60 mph 11.0sec

DOWNTON
Full-race Austin Mini-
Cooper S
¼-mile 15.7sec:
0-60 mph 7.3sec

Drag day for hot Minis . . .

The heat of a hot Mini depends, as with all car performance improvements, on the depth of your pocket. There are quite a few people in the business of making Minis go more quickly, and various ways of using maximum Mini-power other than in Mr Issigonis' space-saving little wonderbox. We managed to persuade a small cross-section of a big market to come to Silverstone's airy spaces for a session of quarter-mile sprinting against a stop-watch recently. Before observant readers start asking questions, it should be explained that our quarter-mile was laid out along Silverstone's Hangar Straight which is not level and that the track was pretty damp, making fast starts difficult. The point, of course, is that testing a collection of similar cars under identical conditions is obviously more interesting.

1. BMC Stage 1 Austin Mini Mk II 998 c.c.

In this particular case fitted by Downton Engineering, the BMC Stage 1 conversion for a normal Mk II Mini is obtainable from any BMC distributor (BMC part number C-AJJ3346). You get a Downton-modified cylinder head, inlet-exhaust manifold and tailpipe and use the standard single carburettor; the kit is just about the most basic there is around and like nearly every Downton one, should improve fuel consumption as well as performance. The engine obviously revved more freely, was in no way made fussy or noisy, remained just as flexible and performed with a clear edge over the original. Price £46 (fitting charge £12).

2. Downton Stage 2 Morris Mini-Cooper 998 c.c.

This conversion was installed in a hack demonstration car which is used by Downton for everything from road-testing engines to dashing from deepest Wiltshire to darkest London. the unit had done a lot of miles yet ran with plenty of zip. It consists of a modified cylinder head and inlet manifold with twin HS2 1¼in. SUs, a standard exhaust system but special long centre manifold and the standard Cooper S final drive ratio, 3.44-to-1 (instead of the Cooper's 3.77-to-1). This was a slightly more sporty car than standard, went a lot quicker—performance figures were taken using an easy 6,500 rpm as the gearchange point—but remained a docile well-behaved little beast. Price £76 2s (fitting charge £14).

3. Oselli Stage 1 Morris Mini-Cooper S 1,275 c.c.

Oselli Engineering are relative newcomers to the tuning business; they are based at Baynards Green near Bicester. Their Stage 1 Cooper S conversion is a not-too-ambitious one whose price will fit the pockets of the majority of prospective customers. It simply involves changes to the cylinder head and manifolds only, these items being polished and slightly re-worked. The result is another adequately tractable and reasonable road-car with no vices, and a better performance.
Price £31 with exchange head.

4. Unipower GT 1,275 c.c.

The Unipower GT is made by Unipower Cars and is that distinctive-looking little mid-engined baby-grand touring car with the space-frame chassis and glass-fibre body that first appeared at the 1965 Racing Car Show. It has its Mini engine and transmission unit arranged east and west; the example that came to Silverstone was fitted with a standard 1,275 Cooper S engine, a Jack Knight five-speed gearbox (extra) and a 2.9-to-1 final drive (as usually offered with the 1275 engine—998 c.c. models come standard with a 3.44 gear set). The Knight box makes rapturous screamy noises and takes a little practice to change quickly. Getting in also takes practice but once you've done it, you're beautifully held by the fine seats. The acceleration is good especially bearing in mind such high gearing.
Price of standard 1,275 car £1,195 in component form.

6. Oselli Mini 7 Racer

First of the two out-and-out track Minis with whose attendance we were honoured (and delighted), and the only 850 there, this totally stripped little canister of a car is the one raced in Mini 7 formula events by Geoff Wilkes. Where interior furniture is concerned it looks as if the householder has defaulted on all his drip-feed commitments except the one for the essential driving seat; there really is very little else. Side windows are Perspex; the highly modified engine has no grille, just an oil cooler to hide its flank from the public gaze. Carburation is through a one-eyed Weber; a Kenlowe fan cools the water. There's little to cool the clutch, which smells thinly during Geoff's long, carefully slipped 6,000 rpm starts. Once fully engaged, the engine bangs its furious way up to 8,000 rpm (9,000 maximum); there's little poke below 6,000 but a surprising amount above. You too can have one of these if you ask—price on application.

5. Downton Touring conversion Austin Mini-Cooper S 1,275 c.c.

Another name by which Downton call this conversion is the "bolt-on" kit. It consists of a modified cylinder head and inlet manifold, the special long-centre exhaust manifold (blowing into the standard exhaust system) and twin 1¼in. H4 SUs. The particular car tested is owned and used by Richard Longman (see below) on the road and has done a hard-worked 15,000 miles. Lacking aircleaners of any sort —carburettor intake trumpets were fitted instead—it was noticeably noisier than usual but not objectionably so. Performance was excellent, bags of torque throughout the range being evident all the way, making an enjoyable car still more enjoyable.
Price £85 (fitting charge £16).

7. Mini-Marcos GT 1,323 c.c.

People look at people in Mini-Marcoi, because the styling is distinctive. You can buy one in various ways, starting with the basic glass-fibre body-chassis shell. It takes the Mini drive unit, suspension etc. and puts them back in the same places, so to speak. Depending on how you get hold of your Mini bits, this is one of the cheapest ways of having yourself a fast and strong-charactered piece of individual transport (the basic shell costs £230). The model tested at Silverstone had a very hot-stuff bored-out power unit which didn't care to run at less than 3,000 rpm. The camshaft was described as "racing", head and manifold were Janspeed, there were twin 1¼in. SUs; it certainly went pretty fast though after four runs it did wet itself when it boiled. Marcos Cars are situated in the very beautiful Wiltshire town of Bradford-on-Avon and unlike a lot of other firms, invite the prospective customer to visit the works. There is a very cordial open air to the organisation which comes out in the sales brochure. It says that "you are welcome to collect" the body/chassis unit yourself, "but make sure you bring an ample supply of rope".

8. Downton full-race-converted Austin Mini-Cooper S 1,293 c.c.

Without in any way denigrating other cars present, this was the car of the day. It is driven by Richard Longman who works for Downton Engineering and is making quite a name for himself in competitions. In this car at the time of writing, he has convincingly won both races entered so far this year, the Mallory Park round of the Redex Saloon Car Championship and the Thruxton Osram Saloon Car Championship race. The engine has a Downton large-valve cylinder head, Downton inlet manifold to suit a 45DCOE Weber, offset rocker gear, a 3.76 "axle", a lot of other changes and produces a claimed 115 bhp; the car weighs around 11¼cwt. Like the Oselli racer, it is completely bare inside. When they shut the door on you, you feel like Royalty at the finale of a cabaret turn—it clatters like a lid slammed down on a dustbin. When Mr Longman starts it up the noise is enormous—the stacatto banging of the exhaust is closer to a variable rate-of-fire machine gun than a petrol engine—the combustion seems entirely external but within a confined space which also contains you. Carefully controlled wheelspin of the fat and shrieking Billy Dunlop Bunter R7 wet weather tyres and we rocketed off. The road-going Minis were doing around 75 mph at the quarter-mile; after practice, this one was doing 91 mph and still hurtling on towards the next corner. The power unit will be homologated for Group 2 in 1970 which is when it will really come into its own. "Quotations" the Downton man said "on request".

Downton Engineering Works Ltd.,
Downton, Nr Salisbury, Wiltshire

Marcos Cars Ltd.,
Greenland Mills,
Bradford-on-Avon, Wiltshire

Oselli Engineering Ltd.,
Baynards Green, Bicester,
Oxfordshire

Unipower Cars,
Ace Works,
Cumberland Avenue,
London NW10

THE MOST DEMON MINI YET

by Brian Foley

PHOTOS: BRIAN FOLEY

ABSOLUTELY UNBEATABLE on the highly competitive Irish Saloon Racing scene is the remarkable turbocharged Complan Mini driven by Alec Poole, 1969 RAC British Saloon Car Champion, and sponsored by Glaxo Laboratories.

The car was built and developed by Poole and mechanical wizard Henry Freemantle, with invaluable technical advice from Syd Enever, chief of MG's Abingdon Experimental Department where work had been progressing on a similar setup for the BL Comps. Dept. before Stokes brought down the Big Axe.

The engine is, basically, a 1.3 Cooper S sporting an 8-port cylinder head with Lucas petrol injection. The turbocharger is a Holset unit which utilises the exhaust gases to pressurise the inlet manifold with fresh air. The actual pressure took some time to figure out so Alec naturally prefers not to disclose the secret psi figures!

The impeller in the turbocharger spins at an incredible 80,000 rpm, and the metal of the housing actually glows red-hot during a long Race! Under bonnet heat is quite a problem, as the inlet manifold reaches a temperature of approximately 110 degrees centigrade, and is now finned for extra cooling. The throttle is wrapped in asbestos to prevent it from melting. A new bonnet is being made with special air ducts to the blower and the manifold, and twin radiators are fitted, one at each side of the engine.

Petrol is fed to the air pressurised inlet manifold by metered injection, and the exhaust gases, mostly spent in spinning the impeller, are discharged through a short pipe — the almost spent exhaust accounts for the flat note of the engine. Exhaust pipe length experiments are currently being carried out, including a very short stub exhaust pipe sticking upwards through the bonnet.

On account of the forced induction the normally high compression ratio has been

If it weren't for those 13in Brabham wheels it would look just like any other Mini Racer, but squatting under that bonnet are 110 deg C of inlet manifold and a red hot turbocharger.

A study in cramming quarts into pint pots. Two rads and a turbocharger all seem to find room somehow.

lowered to 8:1 by use of shorter con-rods. The compression is, of course, considerably boosted by the increased force of the inlet mixture, as with the more conventional supercharger.

Rated at 1.8 litres, the turbo-blown engine produces a claimed 180 bhp at 7500 rpm, and over the winter months Alec reckons there is more to come! An ultimate power output of 200 bhp is on the cards.

In terms of top speed the Complan Mini pulled 8000rpm on the 3.75:1 final drive ratio on the long straight at Phoenix Park, which, with its 13 in wheels, represents a speed of 142 mph. At that speed Alec described the handling as 'twitchy'. Fuel consumption works out at 5 mpg.

The advantages of a turbocharger as against a conventional supercharger are that the former uses up the normally waste exhaust energy, without the power losses associated with the mechanical drive necessary for a supercharger, and without these mechanical drives the turbocharger should be more reliable. Theoretically the supercharger scores in that the power is all there throughout the rev range, but the turbocharging only becomes really effective higher up. This may not present a disadvantage in practice as there is no noticeable power lag with the Complan Mini under acceleration out of slow corners — but then the car is so superior to all other normally aspirated vehicles that one cannot make fair comparisons.

As well as the engine the other major mods include Brabham 13 in mag wheels with 8 in front rims, and 7in rear rims; and the usual rear suspension and subframe are replaced by a British Vita tubular axle, suspended on coil springs and located by two radiius arms and an A-bracket. By the beginning of next season there should be some bigger front disc brakes, too. Braking has been found to be a bit suspect with all that power to keep in check.

The end of the season was capped by an overall win in the BRDC Silverstone big Saloon Car Race, this time with Roger Enever (son of Syd of course) having a one-off drive. Some car to get fastest lap too, at 88.78 mph, against opposition like David Howes' Falcon.

At last—a comfortable Mini

Better equipped additions to the Mini range; civilized cockpit; good torque from big engine, but fussy gearing; bouncy ride with Hydrolastic; adequate performance from Clubman.
Although the new range of Minis is being announced at the Show, they will not be available at the dealers until November 1.

REGULAR READERS will probably have noted our recent waning enthusiasm for Mini motoring. Make no mistake, we'd be the first to acknowledge the brilliance of the Mini concept and that many detail changes have improved it over the years. The trouble, in our view, is that they haven't improved it enough, particularly in comfort and refinement. Our own comparative Group Tests have cogently underlined just how far the Mini had lagged behind several rivals in these two departments.

Ten years and over two million Minis after its introduction—figures that emphasize the car's commercial success regardless of our appraisal from an owner's viewpoint—many of our criticisms have been answered in one go with the introduction of the new long-nose variants (described on page 117) which are a lot more civilized and habitable than any previous Mini. First the Maxi's excellent seats and now these

much improved Mini ones point to a growing awareness within the Austin-Morris division (ex BMC) that seating comfort really matters. Add to this significant changes in the ventilation, furnishings and instrument layout, not to mention the all-synchromesh gearbox introduced earlier and a gearchange that feels much better than before, and the result is a dramatic overall improvement in both creature comforts and drivability. No longer is a long Mini journey something of an endurance test, even though the noise level is still high (particularly so in the 1275 GT) when the engine is extended, and the ride on poor roads as bouncy as ever.

The adoption of a new straight-ahead instrument pod, fresh air vents—something that few other small rivals can boast—and winding windows that eliminate the need to juggle with the old sliding ones, has inevitably meant a reduction in interior stowage space, though what remains is still quite generous. Significantly more comfort for slightly less stowage space seems to us a fair swap.

This report is actually based on three cars, the Clubman in estate and saloon form, and the 1275 GT. The Clubman is perhaps misleadingly named since it is mechanically identical to

PRICES: Mini Clubman £550 plus £170 6s. 11d. equals £720 6s. 11d. Mini Clubman Estate £583 plus £180 8s. 7d. equals £763 8s. 7d. Mini 1275 GT £637 plus £196 18s. 7d. equals £833 18s. 7d. Extras on test cars: Rake adjusting front seats £12 10s. plus £3 16s. 5d. equals £16 6s. 5d. Triplex electrically heated rear window (saloons only) £10 plus £3 1s. 1d. equals £13 1s. 1d.

continued

Mini Clubman and 1275 GT
continued

the ordinary Mini 1000 and therefore has little in the way of sporting aspirations. What it offers is basic 1000 handling, economy (up to 40 m.p.g.) and performance (75 m.p.h., modest acceleration but plenty of low-speed torque) in more luxurious and comfortable surroundings. On the whole, we found it a delightful little car. All three will be in dealers' showrooms from November 1—but not before.

The 1275 GT—not to be confused with the much faster 1275 Cooper S—gets its much better performance from the combination of a more powerful engine (like that in the Morris/Austin 1300) and—surprisingly—lower overall gearing which makes it noisier and fussier than the Clubman. Fatter radial ply tyres also give it better roadholding.

Performance and economy

Three cars, two engines; the one 1,275 c.c. producing 59 b.h.p. (gross) at 5,300 r.p.m., the other 998 c.c. producing 38 b.h.p. at 5,250 r.p.m. Both have single carburetters, and both start with the efficiency customary with an SU instrument—warming up is a smooth snatch-free process since the choke is not a strangler; both are warm and pulling smoothly after a little over a mile, the choke being returnable in rather less.

Separate performance tests on the two 1-litre cars, the saloon and the slightly heavier estate, show them to be pretty evenly matched. Both engines were fairly new and probably tight with under 2,000 miles on the clock so the minor difference in the maximum speeds is academic, 75.4 m.p.h. for the estate and 74.3 m.p.h. for the saloon which, being lighter, was slightly faster over

The new look, with the longer higher nose and the open-plan full width grille; the hotter of the new ones has the side stripes and the GT badge.

the "maximile" and a second faster to 50 m.p.h. However, the 0-50 time of 16.0 s. is still a bit slower than that of competitors like the Hillman Imp and Escort 1100.

The GT is quite a different story. It has a fairly unstressed engine giving about the same output as the 998 c.c. Cooper; so the performance is on a par with that through the gears. However, its extra torque makes it exceptionally lively in top gear, 40-60 m.p.h. taking less than half the time of the present 1-litre cars, and considerably less than the old Cooper too. It compares well against its competitors, with a 0-60 m.p.h. time of 14.2 s.

Performance Mini 1275 GT

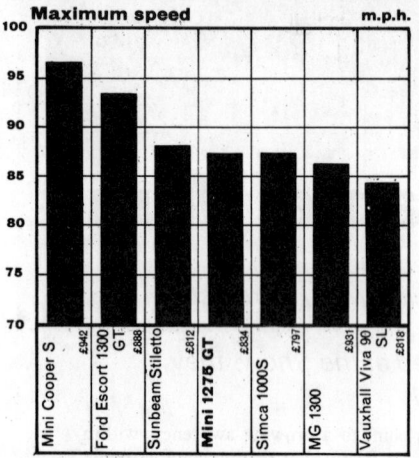

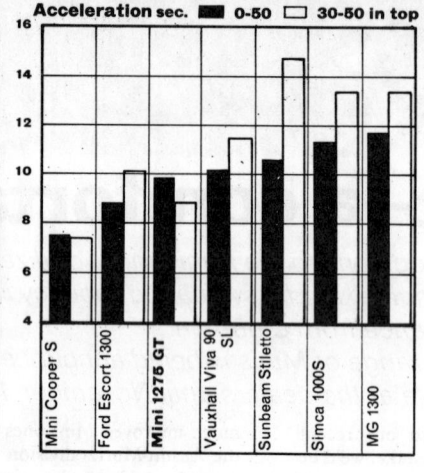

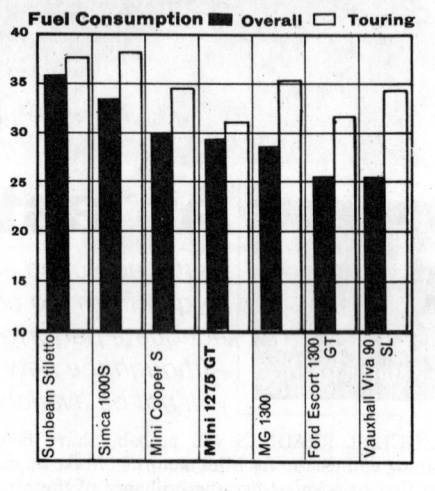

Performance tests carried out by *Motor's* staff at the Motor Industry Research Association proving ground, Lindley.
Test Data: World copyright reserved; no unauthorised reproduction in whole or in part.

Conditions
Weather: Dry with light winds up to 10 m.p.h.
Temperature: 60-66°F.
Barometer: 29.75 in.Hg.
Surface: Dry tarmacadam and concrete.
Fuel: Premium 98 octane (RM) 4 star rating.

Maximum speeds

	m.p.h.	k.p.h.
Mean lap banked circuit	87.5	141
Best one way ¼-mile	90.0	145
3rd gear ⎫	72	116
2nd gear ⎬ at 6,500 r.p.m.	47	76
1st gear ⎭	30	48

"Maximile" speed:
(Timed quarter mile after 1 mile accelerating from rest)

Mean	86.0
Best	88.2

Acceleration Times

m.p.h.	sec.
0-30	4.3
0-40	6.4
0-50	9.8
0-60	14.2
0-70	20.8
0-80	34.3
Standing quarter mile	19.5

m.p.h.	Top sec.	3rd sec.
10-30	—	6.2
20-40	8.4	6.1
30-50	8.8	6.6
40-60	9.9	7.5
50-70	11.8	11.1
60-80	18.0	—

Fuel Consumption
Touring (consumption midway between 30 m.p.h. and maximum less 5% allowance for acceleration)
. . . . 31.9 m.p.g.
Overall . . . . 29.2 m.p.g.
(= 9.6 litres/100km)

Total test distance . . . . 650 miles

Fuel consumptions at steady speed

	m.p.g.
30 m.p.h.	46.4
40	38.6
50	37.7
60	33.0
70	28.8
80	24.5

Speedometer

Indicated	10	20	30	40	50	60	70	80
True	9	18	28	38	48	58	68	78

Distance recorder . . . . accurate

Weight
Kerb weight (unladen with fuel for approximately 50 miles) . . . . 13.2 cwt.
Front/rear distribution 62½/37½
Weight laden as tested . . . . 17.0 cwt.

Just like any other Mini from the back apart from the Mini 1275 GT badge and, on this car, the optional Triplex heated rear window—good value at just over £13.

None of the three engines idled very smoothly, both 1-litre cars shaking sufficiently to set up a vibration in the slender steering column surround. With no rev. counter on the smaller cars we found ourselves changing gear at around 5,500 r.p.m., at which point the engines were running out of breath rather than sounding strained; on the GT, though, the red-lined 6,500 r.p.m. on the rev. counter felt rather faster than the engine really liked or needed, particularly in third gear. This maximum is equivalent to just over 50 m.p.h. in second before valve bounce sets in and it was possible to get 70 m.p.h. in third, against 60 with the others. It wasn't worth delaying a gear change that

long in the standing start accelerations, though. At speeds around 70 m.p.h. all the cars feel quite happy; beyond that the GT begins to sound fussy, and on the test car the gear lever chattered like that of early examples of the remote control linkage.

All three cars were fairly new and possibly capable of returning better economy; even so, both the 1-litres allowed a respectable touring consumption around 40 m.p.g. The steady-speed figures of the GT were very much worse at the lower end of the speed scale (worse than the factory claims), but once the car got into its stride at 70 m.p.h. the consumption was in fact better than that of the smaller cars which struggled to maintain 70 m.p.h. round MIRA. That the small cars still returned 31 m.p.g. when much of the rather low test mileage was at 70 m.p.h. is creditable; the GT did a slightly greater mileage and returned 29.2 m.p.g.

One detail still overlooked is the size of the fuel tank which remains a meagre 5½ gallons, making stops infuriatingly frequent with the GT at its higher average speeds. The estate has a six-gallon tank. All the cars seemed happy on a diet of 4-star fuel.

Transmission

We find it hard to accept that the choice of final drive ratios can be right on both these models; the 1-litre cars have the higher final drive used originally on the 1275S to give more relaxed cruising speeds, but the more powerful GT reverts to the original Mini ratio at 3.65:1 giving good top gear acceleration but buzzy fast cruising. The smaller models stick to the standard Mini ratios, but the GT has a close ratio set which, coupled to a higher rev. limit of 6,500 r.p.m., gives very good overtaking and nearly 50 m.p.h. in second gear against the smaller higher geared cars which run out of breath at around 40 m.p.h. in second. The

continued

Performance Mini Clubman

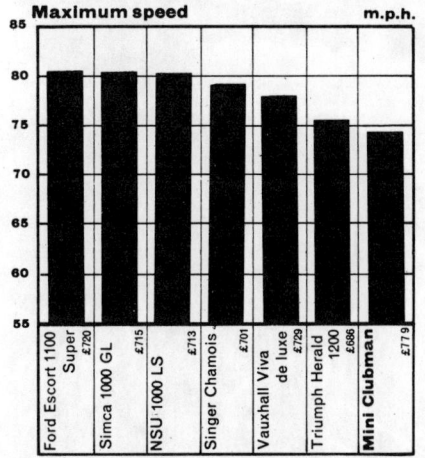

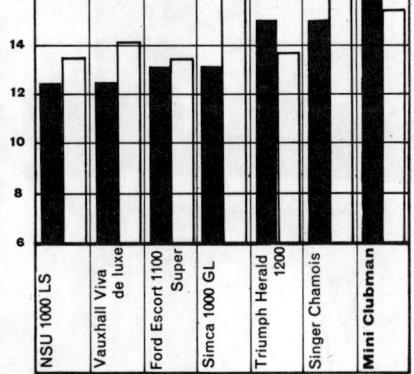

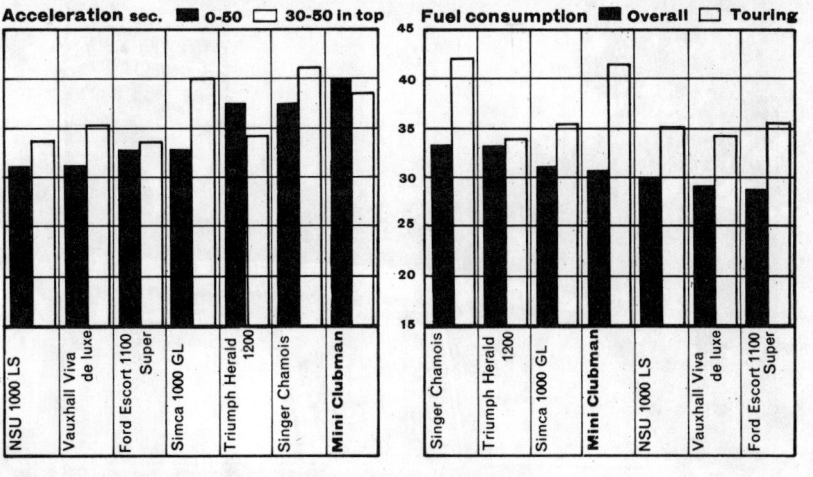

Performance tests carried out by *Motor's* staff at the Motor Industry Research Association proving ground, Lindley.
Test Data: World copyright reserved; no unauthorised reproduction in whole or in part.

Conditions
Weather: Dry with light winds 10 m.p.h.
Temperature: 60-66°F.
Barometer 29.75 in. Hg.
Surface: Dry tarmacadam and concrete.
Fuel: Premium 98 octane (RM) 4-Star rating

Maximum Speeds

	m.p.h.	k.p.h.
Mean lap banked circuit	74.3	120
Best one-way ¼-mile	79.6	128½
3rd gear	63	101½
2nd gear at 5,500 r.p.m.	41	66
1st gear	26	42

"Maximile" speed: (Timed quarter mile after 1 mile accelerating from rest)
Mean 72.0
Best 75.0

Acceleration Times

m.p.h.		sec.
0-30		5.8
0-40		9.7
0-50		16.0
0-60		25.7
Standing quarter mile		22.9

m.p.h.	Top sec.	3rd sec.
10-30	—	8.7
20-40	13.8	8.8
30-50	15.4	10.3
40-60	20.6	15.6

Fuel Consumption
Touring (consumption midway between 30 m.p.h. and maximum less 5% allowance for acceleration)
41.7 m.p.g.
Overall .... 30.7 m.p.g.
(= 9.2 litres/100k)
Total test distance .... 420 miles

Fuel consumptions at steady speed
30 m.p.h. .... 58.8 m.p.g.
40 .... 51.1 m.p.g.
50 .... 45.7
60 .... 37.2
70 .... 28.3

Speedometer

Indicated	10	20	30	40	50	60	70
True	9	18	28	38	48	58	68

Distance recorder .... ½% slow

Weight
Kerb weight (unladen with fuel for approximately 50 miles) .... 12.6 cwt.
Front/rear distribution 62½/37½
Weight laden as tested 16.4

Mini Clubman and 1275 GT

continued

choice on both cars makes them very drivable and one is never caught in a ratio gap. There was a little gear whine on the Clubman but much more on the close ratio set in the GT.

One big improvement since we last tested a Mini is the adoption of synchromesh on bottom gear, a useful and very worthwhile addition. Moreover, the gear-change itself, though still a little stiff (presumably through newness) felt more crisp and positive than before so that gearchanging is certainly not a chore. The smoothness of the clutch and absence of any transmission harshness suggested that there have been other unsung developments in this department since our last Mini test.

Handling and brakes

This is the one heading under which each car is different; both the Clubman and the GT use the same Hydrolastic displacers, but the GT has radial ply tyres on 4½-inch rims as standard while the Clubman retains the original 3½-inch rims. The Estate, however, uses the original "dry" rubber suspension giving it a greater load carrying capacity without the nose being pumped up through the interconnected system.

Of the three, the GT had the best roadholding; on radial tyres, the understeering scrub was much less noticeable and for all normal, but fast, motoring you simply steer through the usual direct Mini rack and pinion and corner with very little roll and a lot of grip. It can be thrown around, too, in typical "dry" Mini fashion to overcome the inherent understeer, and on smooth roads it feels very progressive at most speeds. On bumpy roads

and tight smooth corners, where you have a larger power surplus, it is easy to get the inner front wheel off the deck—easy when accelerating in a straight line over bumps, too.

The failing we liked least was the torque wind-up of the nose under full throttle; to start with it makes a mockery of headlight settings with the large angular change between full power and braking; you get the same effect in a corner, too, which makes the customary lift-off tuck-in a bit less tidy than usual. The amount of tuck-in is probably a little less than on the Clubman since the tyres run at lower slip angles. A passenger prone to travel sickness may well find the perpetual braking/accelerating attitude change a little disconcerting on twisty roads. We would like to see the competition style displacers on the GT, or else stiffer anti-pitch springs.

With less power, the Clubman handles much the same but at lower cornering forces on its cross-ply tyres, and it didn't have such a noticeable attitude change under full throttle; it did suffer from more tyre squeal though. It was interesting to get back to the "dry" estate car after this; its suspension is rather stiffer, so roll is less and it really seemed to handle rather better, being more progressive on the overrun than the Clubman.

On wet roads the radially shod GT clung to the road very well and it was easy to keep the right amount of power down on the road as the slightest change in grip was relayed through the steering wheel; a fair amount of gyroscopic power shake came through, too, under full acceleration in bottom gear. The cross-ply tyred 1-litre cars gripped less effectively but the steering was just as sensitive. All of them have a useful lock with a turning circle well under 30 feet, but the parkability gap required is now a bit longer with the extra inches in the nose.

continued

Specification

Transverse front engine; front wheel drive; independent suspension all round with Hydrolastic displacers on saloons and rubber cones on Estate.

Mini Clubman

Engine

Block material	Cast iron
Head material	Cast iron
Cylinders	4
Cooling system	Water
Bore and stroke	65.58 mm. (2.54 in.) 76.20 mm. (3.00 in.)
Cubic capacity	998 c.c. (60.96 cu. in.)
Main bearings	3
Valves	Pushrod o.h.v.
Compression ratio	8.3:1
Carburetter	HS2 SU
Fuel pump	SU mechanical
Oil filter	Full flow
Max. power	38 b.h.p. at 5,250 r.p.m.
Max. torque	52 lb.ft. at 2,700 r.p.m.

Transmission

Clutch	7¼ in. dia. s.d.p. diaphragm spring
Internal gearbox ratios	
Top gear	1.0
3rd gear	1.433
2nd gear	2.218
1st gear	3.525
Reverse	3.544
Synchromesh	On all ratios
Final drive (type and ratio)	Helical spur, 3.44:1
M.p.h. at 1,000 r.p.m. in:	
Top gear	16.5
3rd gear	11.5
2nd gear	7.4
1st gear	4.7

Chassis and body

Construction	Unitary

Brakes

Type	Drums front and rear
Dimensions	7 in. dia. front, 7 in. dia. rear
Swept areas:	
Front:	66 sq. in.
Rear:	55 sq. in.

Suspension

Tyres	Dunlop C41 5.20-10
Wheels	Pressed steel disc
Rim size	3½J-10

Mini 1275 GT

Engine

Block material	Cast iron
Head material	Cast iron
Cylinders	4
Cooling system	Water
Bore and stroke	70.61mm. (2.78in.) x 81.28mm. (3.20in.)
Cubic capacity	1,275 c.c. (77.81 cu. in.)
Main bearings	3
Valves	Pushrod o.h.v.
Compression ratio	8.8:1
Carburetter	HS4 SU
Fuel pump	SU mechanical
Oil filter	Full flow
Max. power	59 b.h.p. at 5,300 r.p.m.
Max. torque	65½ lb.ft. at 2,550 r.p.m.

Transmission

Clutch	7¼in. dia. s.d.p. diaphragm spring
Internal gearbox ratios	
Top gear	1.00
3rd gear	1.35
2nd gear	2.07
Reverse	3.35
1st gear	3.30
Synchromesh	On all forward ratios
Final drive (type and ratio)	Helical spur 3.65:1
M.p.h. at 1,000 r.p.m. in:	
Top gear	15.0
3rd gear	11.1
2nd gear	7.2
1st gear	4.8

Chassis and body

Construction	Unitary

Brakes

Type	Disc/drum combination
Dimensions	7½ in. dia. disc, 7 in. dia drum
Swept areas:	
Front:	61
Rear:	55

Suspension

Tyres	Dunlop SP68 145-10
Wheels	Rostyle pressed steel
Rim size	4½J-10

Smaller version of the familiar Rostyle wheels standard at 4½J-10 on the GT; 145-10 radials are also standard.

No rev-counter for the Clubman, but the rest is the same as on the GT. Small switch under the right of the facia operates the Triplex rear window.

Left: The Clubman estate; mostly steel with just a bit of simulated wood on the side. Rubber cone suspension on these cars to improve loadability.

Interior of the 1275 GT with the new three-dial instrument nacelle, the fresh air outlets and the new three-spoke steering wheel which is leather rimmed for the 1275.

Virtually unchanged rear seat area of the saloons is still ample for two people unless the passenger reclines his seat to the full (an optional extra on these models, which was much appreciated).

Spare wheel and battery arrangement shown here on the GT which has a stiff cover, while the Clubman has a soft one.

Engine of the Clubman looking just the same as the GT to most people, but the GT has a brake servo. Mechanical fuel pump is on the back of the engine.

Contrast between the Clubman and an early Mini; door hinges are now hidden and the doors themselves are slightly larger for better access. Undivided winding side windows are also evident.

Mini Clubman and 1275 GT
continued

The 1-litre cars still have drum brakes at the front and require heavier pressures than the servo assisted discs on the GT, but they all seemed to work well from their respective potential speeds.

Comfort and controls
The ride on the Hydrolastic saloons is still rather bounay although the seat no longer aggravates the situation; on a bad bump you can actually be jerked in the air—more easily than on the rubber cone springs—but in most conditions the ride is fair for a small car, although it doesn't by any stretch of the imagination provide the claimed big-car feel. However, the interconnected suspension does minimise any sharp pitching which to some people is the most tiresome feature of a bad ride. The "dry" car was far less "crashy" on bad surfaces than we remembered and with a light load ran quite comfortably on all surfaces, although the shorter travel made sharp undulations more noticeable.

General Mini riding comfort has really been improved quite usefully by the new seats; at last they are comfortable, the right shape and in the right place. They now have reasonable thigh support, well shaped if slightly short back rests and the adjustment can accommodate most sizes; the floor brackets have four holes covering a four-inch span if you need the shortest or longest possible settings. The reclining seats fitted to our cars are optional extras; they move through reasonably fine pitch adjustment and for once no-one complained of back-ache—we hope some of the better features of these cars find their way on to the cheaper models.

A further big improvement is the face-level ventilation; pick-ups for this are each side of the front grille and feed to the swivelling eye-ball vents. That on the driver's side works well and can provide a good refreshing blast, but the duct on the passenger's side passes just over the radiator outlet under the wheel arch and picks up a lot of fan noise the moment the vent is at all open.

Fortunately the heater provides a useful output once the car is moving as the single speed booster fan is very noisy; controls for both this and the heater distribution have been rearranged but the pull-out water tap knob is still rather insensitive between hot and cold, and slow to adjust, too. An extra fitted to the saloons was an electrically heated rear screen which demisted well whether the condensation was inside or out.

If one thinks back to the original 850 Mini, the new 1-litre cars are considerably quieter thanks to higher gearing and better insulation generally, although 65 m.p.h. is still about the comfortable limit. Wind noise was seldom obtrusive until you opened the front of the two sliding windows. Now, with fresh air ventilation, you don't need to wind the new windows down; it gets a bit draughty if you do. Still the most prominent noise in the Clubman comes from the engine (mostly from the fan), but it is acceptable by one litre standards. The lower geared GT was less so; more engine noise and an engine that feels rather rougher than that in the larger 1300 combine to make it noisy at 70 m.p.h. when there is also an exhaust harshness. On our test car, the gear lever buzz came in at this speed, so it was altogether quieter at 80 m.p.h. than at 70. Radials on the GT kept the general road noise down, but the cross-ply tyres on the Clubman still roared on coarse surfaces.

Visibility is, of course, unchanged on these Minis and it is still easy to slot them into the usual pint-size holes, more so of course than with the original Minis which had a much poorer lock. On the estate rearward vision is still hampered by the thick centre pillar, so extra mirrors are very useful.

Fittings and furniture
The most obvious feature change inside the car is the new instrument nacelle properly placed in front of the driver's eye. The Clubman has two dials, one speedometer and a combined fuel and water temperature gauge but only a warning light for oil pressure. The GT has a matching rev. counter as well. Most of us liked this arrangement but some felt that the speedometer calibration should be less vague.

The new nacelle has taken up a bit of the large parcel shelf but there is still adequate space on the passenger's side; what has gone, though, with the introduction of winding windows is the vast door pockets—the penalty of progress. The pockets at the ends of the back seat still provide quite generous stowage, however.

There was never a vast amount of space in the Mini boot, but most seem to find it adequate; the Clubman Estate provides the answer for those needing more space as it has a useful luggage area behind the rear seat (thanks partly to its longer wheelbase), but with the squab pulled back and the bask rest dropped forward the flat floor gives a useful luggage area at a convenient loading height.

The interior finish is neat and durable looking with well fitted carpets and a rubber insert for the driver's heels.

Servicing and maintenance
Minis still need servicing every 3,000 miles, but the short service is well within the scope of anyone with a grease gun; the larger bonnet helps accessibility and the different shape no longer catches you on the back of the head as you straighten up. A useful change on these cars is to have a mechanical SU fuel pump instead of the electrical one hidden away among all the dirt and leaves underneath the car.

Longer servicing with oil changes are best left to the dealer. Despite the establishment of Mini as a separate marque the cars will still be sold through the usual Austin and Morris dealerships, of course.   **M**

MAKE: Mini. **MODELS:** Clubman, Clubman Estate and 1275 GT. **MAKERS:** British Leyland (Austin-Morris) Ltd., Longbridge, Birmingham.

Brief Test

A Mouthful of Amals

British Leyland
Special Tuning's
very special
Mini-Cooper S

by Michael Scarlett

Nearly 1,300 c.c., a 12½-compression ratio, an amusing camshaft, an eight-port aluminium cylinder head with an efficient tubular exhaust manifold at the back and four slobbery - trumpet - mouthed Amal carburettors at the front make a 124 bhp Mini-Cooper S which looks most deceitfully innocent.

There is little about the outside appearance of VOH 347J to betray its hairy inside until one plays with the power plus the excellent (standard) Mini-Cooper S handling on a track

FORGETTING the small Special Tuning badge on the side, there is nothing to suggest that this particular demurely pale-green Mini is anything more than a demurely pale green Mini. It does say "Mini Cooper S" on the back, but those excellent little animals are perfectly well behaved if you talk to them nicely.

You have arrived to collect a friend from her home. After the usual 10 minute delay you take her out to the car, politely opening the door and sitting her inside. As it has cooled down meanwhile, you may have to open the bonnet. A short length of twig lives under the driver's seat —you put it there earlier—and because at this particular time you are wearing your best bib and tucker and wish to avoid arriving at the dinner party with fingers smelling of 5-star 100-octane, you use the twig to depress the tickler pins on two of the four carburettors. As an ex-motorcyclist you are familiar with tickling carburettor float chambers but not with the spurt of fuel past each tickler pin that is caused by the pressure of an electric fuel pump.

You close the bonnet, and carry the twig back to the driving seat. Closing the door, you murmur an apology—"Sorry, but the engine's a bit noisy"—then turn the key. With rather more voice than usual, and about five times quicker throttle response the engine fires, sending the rev-counter needle belting round the dial to 5,000 rpm if you aren't careful, after which it settles back to a snuffly 1,400 rpm "tickover".

Further carelessness with the pedals—such as letting the clutch out only a little abruptly at 4,000 rpm in first gear from standstill—produces sharply begun, loudly continuing squealy noises from the frantically spinning front tyres, a tendency to weave and dart about as each tyre takes it in turn trying to grip the road, a further tendency towards smoking, and some pretty rapid acceleration which lands the revcounter at its 7,500 rpm limit incredibly quickly. So you change into 2nd, whereat the

CONTINUED ON PAGE 92

One crosswise cross-flow and most carefully Amaliorated little 1,293 c.c. A-series BMC engine; the engine compartment of an innocent Morris Mini-Cooper S after BL Special Tuning have had a go at it. Sparking plug accessibility isn't as bad as it looks. One virtue of a crossflow Mini head apart from the poke is the greatly improved carburettor accessibility. A quickly removeable grille enables the other major items to be reached. Note, in righthand illustration, the tickler pin on the nearest Amal (under "Amal" sign on barrel of throttle slide) and asbestos bib underneath to prevent over-enthusiastic flooding putting petrol where it shouldn't be

MOTORING PLUS

by Tony Dron

EIGHT PORT MINI

The only giveaway about Special Tuning's very special Mini is the discreet badge stuck on the side of the front wing. Apart from that it could be the faithful old 850 that your aunt goes shopping in.

But there the similarity stops dead. This Mini cuts 19.3 sec. off the 0-80 mph standing start time for a Mini 1275GT—and that's not a misprint.

The heart of this conversion is a 12.5:1 compression ratio, eight-port aluminium cross-flow cylinder head, with four Amal carburetters bolted on the front. Coupled with a full race camshaft the engine gives almost 100bhp per litre, an astonishing figure for a road car when you consider that it's not long since racing engines were unable to perform so well. But the really amazing feature of the Special Tuning Mini is the excellent flexibility: you hardly have to

change gear, yet at 7000 rpm the motor churns out 124bhp.

Of course, this sort of power requires considerable strengthening of the moving parts. The crank is nitrided and special high quality forged pistons are fitted into the cylinders, which have been bored out to give 1293cc. The car we drove had a 4.267:1 final drive, a limited slip differential, and straight cut gears for minimum transmission power loss. The only modification to the Hydrolastic suspension was the fitting of supplementary competition shock absorbers to the front —even the 4½J steel wheels were retained!

When we drove the alloy head Mini the engine was still in the development stage, and although Special Tuning and Amal had got it sorted for full power they were still working hard on cold starting and plug oiling problems. Amals normally have air slides for starting enrichment, but there was no room for these on the Mini engine. Each carburetter was fitted with lawnmower-style, spring loaded buttons for flooding the carbs when the engine was cold, but we found the most reliable coldstart device was a willing accomplice prepared to cover the intakes with her hands.

With this method it always started first time so long as the plugs were clean. Amals have recently changed the angles of certain critical drillings within the carbs, and claim they have overcome any cold starting troubles. These carbs have been specially developed for the cross-flow Mini, so it won't be possible for enterprising amateurs to buy the head and stick some old motor bike-type Amals on it.

The plugs in this engine were extra-small 10mm. ones and the trouble here was that the range of plugs of this size is limited. Special Tuning supplied us with two sets,

one hard (Champion G595) for high speed running, and one soft (Champion G63) for ordinary road use. After a day of driving around town on the hard plugs the engine would start to misfire and it always took a plug change of 10 miles, or high speed driving to clear the motor. Plug changing is easier than it looks: the thing is to loosen them off with an ordinary plug spanner and then push a piece of small bore water or oil hose over the plugs and spin them out one by one. But with an aluminium head that costs £230 you want to keep plug changes down to a minimum: it would be very easy to strip the threads.

We saved a new set of hard plugs for the MIRA testing, and as we changed them we noticed one of the plug connectors had chafed and was partially shorting out. This may well have exacerbated the oiling-up problem, and as a result Special Tuning are now using a different type of plug connector.

The temptation to blast round the countryside at full bore in this car was absolutely irresistible, an experience that no boy racer should miss. Of course, we're not saying you'll find boy racers at *Motor*: it's just that nobody with the slightest enthusiasm for

The front of a Mini engine never looked like this before

driving could possibly step into VOH 347J and drive off without getting that leadfoot feeling within 50 yards.

There's just no way to describe the performance of the car—it was even quicker off the mark than the Downton 1275GT, Tiger Tim, that was tested for our Tuning Supplement earlier this year. We reckon it might have been even quicker with a slightly higher final drive ratio—with the very low 4.267:1 the sort of gear you might fit for a hill-climb, there was so much gearchanging to be done at low speeds that it can't have helped the getaway. This very low ratio, which gave roughly 39, 58 and 80mph in the intermediates, was a hindrance in normal use since the car was rev limited even in top, at exactly 100mph. The maximum permitted 7500 revs could be reached with ease and it would be all too easy to over-rev the car in top gear, if you took your eyes off the tachometer.

We regarded Tiger Tim as being the most sophisticated tuned Mini we have ever tested and we stick by that assessment. Special Tuning's incredible device is aimed at a different market entirely: if Tiger Tim was the most sophisticated then the alloy head Mini joins the ranks as the most successful compromise we have yet found for the man who wants to use his road car in competitions and win. Tiger Tim was pretty civilized yet very quick indeed. The alloy head machine was incredibly quick but must rank as the noisiest car (for the occupants, that is) we have

---

Car: BLMC Mini Cooper 'S'
Tuner: British Leyland Special Tuning Dept., Abingdon, Berks.
Telephone: Abingdon 251
Conversion: Aluminium cross-flow head with four Amal carburetters, full race camshaft, block bored out to 1293cc, lower final drive and close ratio straight cut gearbox, competition dampers at front.

	£
Balancing rotating assy.	10.00
Boring out block to 1293cc.	5.00
Pistons (Set of 4)	25.00
Head (8 Port Aluminium)	230.00
Head gasket	1.75
Camshaft	25.00
Sprocket (lightened)	7.50
Spacer—valve rocker	0.33
Screw—tappet adjusting	0.28
Lightened tappet	0.43
Exhaust manifold	20.00
Manifold gasket	1.05
Lightweight flywheel	29.05
Clutch assy	2.50
Clutch plate	3.66
Locking plate crankshaft pulley	0.11
Main bearing nut set	0.25
Primary/Idler gear straight cut	40.00
Special inlet manifold	60.00
4 Amal carburetters	60.00
Pick-up pipe—oil pump	1.65
Pulley—dynamo	0.80
Fan Belt	0.45
G59R Sparking plugs	0.75
High tension kit	4.75
Close ratio straight cut gears	32.00
Limited slip diff	42.00
Total	**£545.11**

	Special Tuning Mini 1300	Downton 1275GT	Standard 1275GT
Maximum speed			
Lap	100.0*	104.3	87.5
Best ¼ Mile	100.0*	107.1	90.0
Mean Maximile	100.0*	102.2	86.0

Acceleration			
mph			
0–30	3.5	3.1	4.3
0–40	4.9	4.4	6.4
0–50	6.5	6.4	9.8
0–60	8.6	8.4	14.2
0–70	11.3	12.0	20.8
0–80	15.0	15.9	34.3
0–90	19.7	21.1	—
0–100	—	40.3	—
Standing ¼ Mile	16.4	16.5	19.5
In Top			
mph			
20–40	7.2	11.5	8.4
30–50	5.5	11.2	8.8
40–60	5.6	9.7	9.9
50–70	5.9	9.8	11.8
60–80	6.7	11.7	18.0
70–90	8.4	15.8	—
In Third			
mph			
10–30	5.7	7.9	6.2
20–40	4.7	7.5	6.1
30–50	4.2	6.8	6.6
40–60	4.3	6.0	7.5
50–70	4.7	6.2	11.1
60–80	—	7.4	—
Fuel consumption			
Steady mph			
30	27.4	45.4	46.4
40	31.6	46.5	38.6
50	29.1	44.6	37.7
60	26.7	39.5	33.0
70	24.7	36.6	28.8
80	22.4	33.6	24.5
90	15.6	28.6	—
Overall	19.2	29.0	29.2
Touring	25.7	35.5	31.9

* rev limited

Just any old Mini from the outside . . . Four Amal carbs, a cross-flow aluminium head and a full race cam make an extra tractable Mini turning out 124 bhp at 7000 rpm

ever driven. One of our staff actually wore the rubber earplugs he has for full bore rifle shooting at Bisley to drive the Special Tuning Mini. At low speeds in top the straight cut gears can be heard chattering and whining away, but the moment you let the horses go the transmission noises are drowned and you know there's a full race motor on the end of your boot.

When you use full power in second and third, even top on some roads, you can feel the little wheels fighting for grip and tugging the steering first one way and then another as you tear over bumps in the road. It's all very controllable, although one of our drivers felt that the front end was just a little under-damped despite the competition shock absorbers. In fact the handling was a joy: no doubt the roadholding would be a bit better with wider wheels, but it was still very good with the 4½Js and the car could be chucked about in an alarming fashion (for the passengers anyway) without a hint of adrenalin creeping into the driver's stomach. You just sling it sideways and turn on the power to pull you out of the corner—you can rely on the limited slip diff to stop the inside wheel spinning and so force every ounce of power down on to the road.

The brakes, too, were very good despite the lack of a servo. With a separate inlet pipe for each cylinder there's not much hope of connecting up a servo and in any case, Special Tuning reckon it would upset the finely balanced carburation. Not one of our drivers thought the brakes were even noticeably heavy without one—in fact most of us preferred having to apply a bit of pedal pressure.

Special Tuning's objective in producing the alloy-head is to keep the Mini up at the front in club competitions. It looks as if they've made the grade.

GO-CHEAPER WITH ZENITH?

It's not all go-faster in Motoring Plus, you know. We had go-comfortabler (I insist there is such a word) a couple of weeks back with University Motors, and now we're trying go-cheaper with Zenith.

"If you operate an 1100, 1300, 1500, or 1600cc Escort, Capri or Cortina, we can supply a sparkling new economy conversion which could reduce your petrol bills by up to £25 pa. For further details, complete and return coupon to Zenith NOW!" So says the ad., and as it happens we do run a Capri 1600, a venerable machine whose first 50,000 miles of faithful service to *Motor* was reported two weeks ago. We sent off for details from Zenith at Honeypot Lane, Stanmore, Middx., to see if their conversion could improve on the remarkable 27.0 mpg regularly turned in by our staff Capri. Obviously it would provide a stern test for the magic device from Honeypot Lane.

The Zenith conversion, complete with air cleaner, gaskets, choke cable, and all the bits and pieces required for the changeover costs £13, or £11 if you exchange it for your old Autolite carburetter. It can be no coincidence that an Autolite replacement costs £11 from FoMoCo, though no exchange service is available.

We cover about 16,000 miles each year in the Capri which must be about average for this kind of car. At 27.0 mpg we get through around 593 gallons so to save £25 this must be cut by around 70 gallons to 523 gallons. In other words we should get about 30.5 mpg to equal Zenith's maximum claim . . . .

Having verified the Capri's overall fuel consumption of 27 mpg we arrived at MIRA and carried out all the checks on the standard carburetter—acceleration in the gears and constant speed consumption with the Petrometa. Next job was to install the replacement Zenith: to save time we had read the instructions earlier and so we got on with the task as follows:

12.40 pm Job begun.

12.50 pm Old carb removed.

12.55 pm New carb bolted on. We noted that with the thicker gasket as supplied by Zenith there were still two turns of the threads exposed inside the nuts. Okay for our purposes, we decided.

1.00 pm Despite incorrect instructions about replacing vacuum advance line common sense got us over the problem easily enough. The fuel line went back on with *just* enough length to spare. Inadequate instructions for refitting throttle.

1.20 pm Engine running, (started first time in fact), tickover and idling mixture screw adjusted.

The standard Autolite carburetter, above, and the replacement Zenith, right

The Zenith looked very smart and the job was really very easy and snag free. We could do the job in less than 20 minutes next time.

We pressed on with the test and recorded the results shown in the panel. We found that the car with the Zenith was a little slower at the bottom end and rather quicker at the top end in fourth gear, though really there was not much to choose between the two in terms of performance. Although the Zenith is intended for economy, not for speed, it gave noticeably more pull when flooring the throttle at 80 mph. The Autolite gave smoother running at low revs, however. We also noticed considerably more intake roar with the Zenith, due to the different air cleaner. We faced the open side of the cleaner to the front as instructed, but the noise level was quite high even so.

As for economy, our touring fuel consumption figures, which are calculated from the Petrometa results, are identical for both carbs for all practical purposes. Here again, the Autolite scores well at low speeds and the Zenith wins at the extreme top end.

Unfortunately we were unable to carry out a long term overall check, but the Petrometa results lead us to believe that the Zenith will probably return similar figures to the Autolite if the engine is working properly. There was certainly no problem involved in starting and no rough running when we installed the Zenith at MIRA.

Would we recommend the Zenith? This is a hard question to answer, especially as we found that the Zenith was no more economical than our Autolite. Autolite carburetters are turned out at the rate of more than 5000 per day for Ford cars all over Europe. Ford used to fit Zenith carbs on many of their saloon cars until they decided to make their own. Both carbs are clearly well engineered products, though Zenith claim that their production methods ensure more consistent standards of quality. Certainly the Zenith looks better made and it has replaceable jets instead of mere drillings in the casting.

Both units achieve the same object, but by a different means. The Zenith is set to run rich for maximum power when the throttle is opened wide but employs an economy device. When the throttle is eased off at cruising speeds a diaphragm, sensitive to manifold depression, opens an air bleed which weakens the mixture and cuts down consumption. The Autolite is set to run weak for economy, but in this case manifold depression is used to operate a "power valve", a device which acts as an adjunct to the main jet and provides a continuous stream of fuel when the throttle is floored for maximum power. Both units are equipped with accelerator pumps for rapid pick up. Zenith reckon their way is more economical, while Ford say their power valve system is more sophisticated and achieves a more consistent air/fuel mixture. Results seem to show that in practical terms there is not much difference in performance or economy.

Ford add that emission regulations have demanded greater accuracy in the manufacture of carburetters and that Autolite are now manufactured to closer tolerances than ever. Our opinion is that if your Autolite is working properly, leave well alone. If not, the Zenith conversion is worth serious consideration.

Acceleration In top mph	Capri 1600 with standard Autolite carburetter	Capri 1600 with Zenith replacement carburetter
20-40	9.6	10.3
30-50	9.1	9.5
40-60	10.1	9.9
50-70	12.1	12.2
60-80	17.6	17.1
In Third mph		
10-30	7.5	7.7
20-40	6.3	6.4
30-50	6.7	6.6
40-60	7.9	7.9
Fuel consumption Steady mph		
30	45.6	43.9
40	42.2	41.8
50	40.0	39.0
60	28.9	28.7
70	31.0	30.4
80	23.3	25.4
Overall	27.0	25.0*
Touring	27.4	27.3
*see text		

Guess where this month's Home Brew comes from. Good old Poole once again, bringing the total to four Brewings from that green and watery township. Makes me wonder though; it doesn't seem natural that so much good stuff should come out of those few square miles, and suggests something mysterious. I have this feeling that Poole-modified cars aren't built — they're grown. Yes, that's it, I should have realised sooner! Wander into a gardening shop in the town centre and I bet you find little seed packets labelled 'Dwarf Variety Minisprint' or 'Hybrid Anglia TR4' nestling among the packs of Morning Glory (the trip you can eat between meals) and Viola Tricolor. Just think of the planting instructions — like, 'support on A-brackets and water with a biochem mix of Wynns and GTX. If spacing greater than 1½in use RAC approved arch extensions.'

Not that anyone down there will admit it of course. For instance, Clem and Marty Orman swear blind that their Minisprint was built with their own fair hands — but then, a good gardener never gives away his secrets, so I'll just have to tell the story their way.

The Orman team is a father/son duo, Clem being pére and Marty fils. Clem, a printing engineer, has been mechanically active since way back, having raced grass-track motorcycles before the war. He then changed to four wheels and played around with such delights as a tweaked 1937 Y-type Ford. However, all his more recent work has been devoted to Minis, a love for the breed being formed with his first Mini in '63. Since then he and Marty (who's a design draughtsman) have remained faithful, and never so much as eyed up other automotive talent.

In all the Orman family has owned six Mins, the best to their way of thinking being an evil little black box camouflaged to look like a standard 850 but sporting a sneaky full-house 998 mill. It had been rolled twice and, in Clem's words, 'was a really murderous weapon'.

The idea of a Minisprint was conceived in 1971, with the Ormans fancying a variation on the ordinary Min theme. However, Sprint shells are rather thin on the ground (although Goochie has one he'll sell for a sum not unadjacent to 1-million pounds, or complete with 1000cc engine and etceteras for £350) so what to do? Well, the original Sprint was designed by Neville Trickett of Siva fame who builds, amongst other things, those vintage replicas on Ford Pop and VW chassis, so it was to Nev that Clem turned. Mr T wasn't too keen on getting involved in Mini-chopping again, but was eventually persuaded to cut the bodywork and tack-weld it together, leaving the final welding to the Ormans.

The method of cutting was very clever, enabling the use of a full-size but raked back screen. A number of Minisprints aren't the pukka thing because only the roof has been lowered. Not so the Orman car, which has had the body cleverly lowered all round by separating the body from the floor pan and removing metal from the base of each panel. Likewise with the doors, but the front is a glassfibre one cut by Clem. All sounds quite simple, doesn't it? Bet it's not when you get down to it though. The front screen pillars were cut and raked back after metal had been removed above the doors — but how the hell the roof was replaced without it

requiring sectioning I'll never know.

Nearly one-hundred welding rods were used to stick all the bits back together and de-seam the shell. The dash panel was removed and replaced by an ally/leather-cloth panel built round two lengths of 1in square tubing which run from either side of the front bulkhead.

A Minivan fuel tank now completely fills the boot, so a frame for the spare wheel has been constructed from Dexion and bolted to the rear bulkhead inside the passenger compartment and alongside the battery.

The 998 Cooper engine has been fully rebuilt to (for want of a better term) half-race spec by Dale Clements at Taurus Engineering. All the entrails have been lightened and balanced and the bottom end treated to a steel main-bearing strap. A Taurus 211 cam (roughly equivalent to a works 731) operates valve gear which has been modded as follows: forged steel rockers, steel pillars, Hidural valve guides and large Nimonic valves. This little lot is nicely mounted in a 12G295 head casting which has been fully gas-flowed, ported and skimmed to give a comp ratio of 10:1. The timing gear is Duplex and both water and oil pumps have been modified to give greater capacity.

At the moment the carbs are a pair of 1¼in Strombergs; Clem's opinion of these can be summed up in saying that by the time you read this they will have been replaced by a brace of SUs. Incidentally, Clem is highly critical of all devices which are claimed to make carb setting easier, but

rates the Colortune plug very highly; just thought you might like to know.

The Stroms are (were?) on an aluminium manifold from Minisport, and the exhaust system is made up of a Janspeed LCB manifold and a Sprite MkI silencer box.

Cooling is carried out by a Mini rad, mounted at the front and on its side in true Min-racer fashion, and coupled to a header tank made from an 1100 heater expansion tank. Alongside the rad is an oil cooler.

Transmission is by a close ratio helical cut four-speed synchro box, lightened flywheel, diaphragm clutch and 3.75:1 f.d.

Now the bits at the bottom: the front end was originally fitted with Cooper discs, but as their stopping power proved to be notable only in its absence they have been replaced by a Lockheed servo and single-leading-shoe drums with AM4 linings — an improvement of about 100%, reckons Clem. Those Cooper discs really were disgusting, weren't they? Front suspension is lowered, and damped by shorter uprated Armstrongs. Clem lengthened the lower wishbones himself to give about 1½° of negative camber on the 5J steels and Goodyear GPs.

The standard drums and linings have been retained at the back and the shockers are Lockheeds. The rear wheels now have 1° negative camber and no toe-in after some adjustment of the swing arms.

And that, together with all the usual bits like leather rim steering wheel and full instruments, is about that. A neat little Minisprint planted and cultivated in Poole by the family Orman. ■

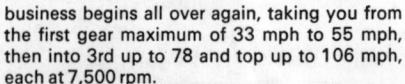
CONTINUED FROM PAGE 87

business begins all over again, taking you from the first gear maximum of 33 mph to 55 mph, then into 3rd up to 78 and top up to 106 mph, each at 7,500 rpm.

If the lady weighs 16-odd stone like the man on the stopwatch when we took acceleration figures at MIRA, you will only get to 50 mph in 6.5sec, 60 in 9.1, the ¼-mile in 16.8 and 100 in 24sec. Although she knows little about motor cars, she usually realises that homely pale green Minis don't as a rule go like that, and starts enquiries. That is assuming that she has got over her first reaction to this incredible machine, which in my case is, I am afraid, near-helpless laughter. There is something immensely comic about the magnificent but hamster-like hunched-up little Mini when you can make it go like this.

How it's done

One reason why the car looks so ordinary is also the main reason for harking back to the days of Freddie Dixon by fitting four Amal 33mm concentric type carburettors instead of the more familiar wear for British Leyland Special Tuning eight-port aluminium cross-flow cylinder heads. Two Weber twin choke carburettors mean alterations to the roof of a standard Mini-Cooper S powerhouse; you have to let-in a dormer window with wire-mesh instead of glass, which is contrary to Group 2 bodywork regulations.

These are not ordinary motorcycle concentric Amals, but specially re-machined ones designed to cope with the high sideways surges encountered in cars—especially sideways Minis—unlike motorbikes. Complete with inlet mani-

minded, only putting one big foot to floor and those four little throttle slides to their roofs just before 2,600 rpm where the "full race" camshaft begins to work.

That is where the noise starts, the wet-mouthed Amal quartet—there's a slight smell of petrol whenever you drive it hard—making a nagging growl like a dog worrying a rat. This becomes a steady roar as the revs rise, drowning most of the transmission noises; you can "cruise" it at 6,000 rpm in top (85 mph) provided you don't want to talk to anyone, though one doubts if "cruise" is quite the right verb, implying restful long-legged high speed. Before we took delivery of the car a 4.27 final drive had been replaced with a 3.9-to-1 set, which greatly improves the car for use on the road, but at 14.2 mph per 1,000 rpm in top it is still under-geared. As the point of the car is possible use on both road in and club competitions, this objection is immaterial.

The engine itself has a fully balanced crankshaft / flywheel / clutch assembly, with lightened flywheel and competition clutch. Bored out by 0.020 in. to give 1,293 c.c., it uses special forged pistons and a 12½-to-1 compression ratio, and with camshaft part no. C-AEG 636, lightened tappets, Champion G59R 10 mm sparking plugs, a nice tubular exhaust manifold and those neat little carburettors gives approximately 124 bhp at 7,000 rpm. Fuel consumption is as heavy as you'd expect—varying between 19 and 27 mpg. Other changes are limited to the fitting of competition front dampers on the Hydrolastic suspensions with, as usual on the Cooper S, 4½J wheel trims. The normal remarks apply to the springing—bouncy despite the dampers, and absolutely stable and precise. We couldn't resist playing with it at MIRA, belting into an open bend and throwing the car sideways, putting power *on* in the usual back-to-front front-wheel-drive Mini fashion to straighten it out when one felt like it. MIRA's unusually high-grip surfaces meant a bit of wheel lifting, but nothing to worry about in this most enjoyable of all Minis. Without the customary vacuum servo—left off because of the awkwardness of taking a vacuum bleed off the separate manifold pipes—the brakes needed a very good shove, but you get used to that.

Certainly, at £530 not the cheapest of quick Minis, but probably the fastest (depending on how it's geared) and quite certainly the most fun of any Mini we have driven. It must appeal strongly to the club man who has to use his competition car on the road. □

Performance Check

Maximum speeds

| Gear | mph | | kph | | rpm | |
	R/T	Amal Mini	R/T	Amal Mini	R/T	Amal Mini
Top (mean)	96	106	155	171	6,000	7,500
(best)	98	106	157	171	6,050	7,500
3rd	74	78	119	126	6,250	7,500
2nd	55	55	88	88	6,500	7,500
1st	33	33	53	53	6,500	7,500

Standing ¼-mile, R/T: 18.4 sec. 75 mph
Amal Mini: 16.8 sec. 89 mph
Standing — R/T: — sec.
kilometre — Amal Mini : 30.5 sec. 107 mph

Acceleration,

R/T:	3.5	6.0	8.2	11.2	15.4	23.4	34.7	—
Amal Mini:	3.3	4.9	6.5	9.1	11.5	14.7	18.4	24.0
Time in seconds	0							
True speed mph	30	40	50	60	70	80	90	100
Indicated speed MPH, **R/T:**	32	42	52	62	72	82	92	—
Indicated speed MPH, **Amal Mini**	31	42	52	62	73	84	95	107

Speed range, Gear Ratios and Time in seconds

| MPH | Top | | 3rd | | 2nd | |
	R/T (3.44)	Amal Mini (3.94)	R/T (4.67)	Amal Mini (5.34)	R/T (6.60)	Amal Mini (7.54)
0-20						
10-30	8.2	—	5.5	—	4.0	3.9
20-40	7.5	—	5.4	5.5	3.8	2.7
30-50	7.5	7.9	5.4	4.7	4.4	3.0
40-60	8.3	7.0	5.4	4.5	—	—
50-70	9.4	7.4	7.5	4.4	—	—
60-80	12.3	6.9	—	—	—	—
70-90	17.5	9.7	—	—	—	—
80-100	—	9.8	—	—	—	—

Fuel Consumption

Overall mpg, **R/T:** 28.5 mpg (9.9 litres-100km)
Amal Mini: 19.9 mpg (14.2 litres-100km)
NOTE: "R/T" denotes performance figures for Morris Mini-Cooper 1275S tested in **AUTOCAR** of 14 August 1964.

Normally he is a fussy little fellow who bustles not specially very effectively about his business, so those naughty motor-car surgeons at Abingdon carry out a tricky, expensive, and very comprehensive bit of organ-swapping on the dear little man which is like putting a bomb under a bureaucrat; suddenly a mouse that roars, he still scampers but with immense excitement and commensurate speed, quiveringly cross, absolutely furious with everyone, like a small boy who has been laughed at, bursting to show anybody and everybody who dares to point their car down a road in the same direction as himself what a tremendously important mouse he is.

He does that by doing everything in short, almost staccato bursts. Obeying speed limits in towns, it is hard to resist the temptation to use the clutch carefully and accelerate as hard as possible away from the lights just short of noisy wheelspin up to 30 mph in 1st, then plop it into top and watch the others catch up, many most clearly curious, as the car looks so straightforward from outside.

folds and a multi-levered throttle linkage that looks like a Meccano model of the controls of a fairly major railway signal-box, the carburettor kit costs a quite reasonable £60 (reasonable when you compare it with other induction systems for this head). The linkage works very nicely, though we did suffer one carburettor cable failure, perhaps due to the rather tight corner it has to negotiate leaving the top of the throttle slide housing. That happened at MIRA, and we must thank Messrs Norton Villiers Ltd for presenting us with the beautifully made twin-throttle cable for a Norton Commando which we guiltily murdered with pliers to provide a replacement cable. Stripping the No. 3 Amal reminded us what an incredibly simple instrument that carburettor is. Apart from the lack of a cold-start enrichment device—not very important on such a car anyway—the set-up works wonderfully, although, remembering motorbike experience we did find it best not to slam open the throttles to flat-out at low revs; gradually opening them one could potter away from 1,500 rpm if one was so oddly

Parts List	Part No.	£
Pistons	C-AJJ 3377	29.00
Boring out cylinder block	—	5.00
Head 8 Port Aluminium	C-AJJ 4064	230.00
Head Gasket	C-AHT 188	1.75
Camshaft	C-AEG 636	25.00
Sprocket lightened	C-AEG 578	8.00
Spacer – valve rocket (ea.)	C-AEG 392	0.38
Screw – tappet adjusting (ea.)	C-AEA 692	0.28
Lightened tappet (ea.)	C-AEG 579	0.43
Exhaust Manifold	C-AHT 343	20.00
Manifold gasket	C-AHT 380	1.05
Lightened flywheel	C-AEG 619	29.50
Clutch assy	C-AEG 481	2.50
Clutch plate	C-22G 247	3.70
Locking plate crankshaft pulley	C-AHT 146	0.11
Main bearing nut set	C-AJJ 4013	0.25
Balanced rotating assembly	—	10.00
4 Amal carburettor kit complete with Manifolds	C-AJJ 4083	60.00
Pick-up pipe – oil pump	C-AHT 54	1.65
Pulley – dynamo	C-AEA 535	0.80
Fan Belt	C-AEA 756	0.50
G59R Sparking plugs (ea.)	C-AHT 435	0.80
High tension kit	C-AJJ 4010	4.75
Close ration straight-cut gears	C-AJJ 4014	32.00
Limited slip diff	C-AJJ 3387	45.00
3.9 Final drive wheel and pinion		13.00
		£530.18

The Cooper S was Leyland's best...
LITTLE BIG HORN CAR

The Cooper S is already well on the way to becoming a mid-sixties classic. AJ van Loon, who owned one "way back when . . .", drives one again and reminisces on its performance past . . .

COOPER S

"When you are reaching the limits of adhesion, things start twisting and the door frame starts flapping so much you can see daylight through the gap."

Total immersion, complete in-osculation, an intercourse of mind and machine, the relationship between an enthusiastic driver and a Ferrari Dino V6 is possible with no other car I can think of.

Except one.

A Cooper S.

Though the Ferrari and Cooper operate on vastly different planes of economic and social wellbeing and are worlds apart in design and mechanical sophistication, the relationship between both cars and their drivers is amazingly close.

It's difficult to explain. A Cooper S can't match a Ferrari in performance or road manners, but like the Fazz it is like an extension of your arms and feet. A mechanical shoebox that absorbs you, vibrates you, assaults your senses with the roar of carburettor intakes, and is ready to do almost everything at your command.

I owned a Cooper S once, and though it was frighteningly expensive to maintain, it was the car I enjoyed owning more than any other before, or since.

There is a corner, an open bend really, on the road that leads from One Tree Hill to Elizabeth, that sweeps left over a slight crest and *down out of farmlands into the built-up area of the South Australian satellite city. Back in the days when it was still rarely used it was possible, with a certain amount of courage and optimism, to take that bend at a true 160 km/h (100 mph) flat chat in the old Cooper S, with Michelins clinging desperately and an unwavering line highly necessary. Then you'd really have to get on the brakes to wash off some 32 km/h (20 mph) for a following right hander and the 72 km/h (45 mph) speed zone that came up after that. I only managed it two or three times but the feeling that swept through me as the heartbeats returned to normal was one of satisfaction, of accord between a driver and his car.*

Recently I made re-acquaintance with a Cooper S. It was an ex-NSW *Basics only — the Cooper S was not overendowed with instruments, even a tach had to be fitted by the owner, an oil pressure gauge was BMC's only concession for the questioning owner.*

Police car which Sydney racer David Clement offered to let my drive. Clement is a Mini specialist and operates a reasonably successful workshop and sales business in suburban Brookvale. He's been at it for about four years and has become known as THE man on Sydney's North Shore for quality Mini repairs.

The dark blue Cooper S I drove away in differed from my own car in that it had been fitted with the police performance kit which consists of 38 mm (1½ inch) SU carburettors and a slightly modified

Left:
Square and plain, a shape totally removed from any sporting pretence, but underneath a character that was performance to the core — the Cooper S, with nose-up attitude indicating a rapid getaway.

head. Also the steering wheel had been replaced by a dished leather rimmed affair, where the one on my car had been a woodrim — a kind which was very "in" at the time.

Sitting behind the wheel the memories flowed back to me: the lumpy idle, windscreen wipers that lift off at anything over 96 km/h (60 mph), and the all-enveloping roar that drowns out everything else at 130 km/h (80 mph).

But how did I ever get comfortable in a Cooper S? Lowering myself into Clement's car I became very aware of the kitchen chair type of seating, the high, flat steering wheel and the very personal nature of the mini-bin.

I was also made aware again of personal vulnerability in the car. Of just one skin of metal protecting you from intrusions through the side, and of what we used to call "cardboard box" type of construction — where if one side was punched in, the rest would collapse around it. But in those days, before the advent of Australian Design Rules on front and side impact protection, we lived with that thought and the blissfully optimistic idea that if one body was "used up" we'd just throw it away and start again.

Rude jolts through the bottom of that "bucket" seat also reacquainted me with the uncompromising ride of the S. A bouncy, jolting progress that would lift the whole car off the ground if you tried cruising at more than 120 km/h (75 mph) on that patched up piece of broken bitumen that passed as the Sturt Highway on the Hay side of Balranald. The old S covered the 132 km (82 miles) between those two towns in 58 minutes one time — 'and used up 18 litres (four gallons) of petrol doing it. Cooper Ss became a little thirsty if they were being booted along.

It is surprising now to realise how many people were willing to put up with considerable personal discomfort for the sake of performance — but around the mid sixties the Cooper S was THE car to have. After all Brian Foley and Peter Manton were giving everything else the big hurry-up on the race tracks and people like Evan Green and Bob Holden were showing that the lowslung little per-

The guts of the matter, 1275 cm³, twin SUs, and enough plumbing to service a house, in an engine bay the size of a rubbish bin.

Below:
Twin fuel tanks status symbol — "Fill 'er up — both sides" being the

off-the-shoulder direction at pump jockeys. Though they reduced boot space to an overnight bag, a gallon of oil and a pint of brake fluid, the twin tanks were a good thing and allowed you to go at least 320 km (200 miles) between fuel stops.

formers would take a lot of punishment in rallies.

Looking under the bonnet of the Clement car also brought back memories of hours spent working in spaces where a spider couldn't crawl. There's one thing for sure, working on a Cooper S taught you self discipline and cured impatience.

Easter time in South Australia usually includes two days of motor sport with a hillclimb on the Saturday and motor races on the Monday. So it came as an unpleasant surprise to find that the bottom heater hose had developed a major leak on Good Friday night. On most cars changing a bottom radiator hose would present no problems, but a Cooper S . . . ?

It took about half an hour to get the offending part out — the hot,

soft, ruptured hose tore relatively easily and with no more than a couple of skinned knuckles was brought up to the light of lamps. Then came the job of fitting the new, stiff hose in place where it was impossible to see what you were doing if you were trying to get the hose on — and it was impossible to get the hose on if you wanted to see what you were doing. Even liberal quantities of grease around the radiator outlet failed to help and it finally became necessary to remove the top radiator hose, the fan shield, the fan, water pump pulley, and radiator to fit the hose onto the bottom. Then the whole thing had to be reassembled.

It took me from 9 pm to after midnight before I had that damned car running again . . .

Continued

LITTLE BIG HORN CAR
Continued

As long as you are able to keep moving the Cooper S excels in city traffic, the instant acceleration and ultra direct steering making the car more manoeuvrable than anything else on four wheels. Similarly on twisting mountain roads the Cooper S shines — you can go suicidally deep into a corner, jump on the brakes, and throw it into a corner with a power on/power off series of slides to help you around if you are going too fast to power through all the way.

Great gobs of power in second gear will induce great gobs of wheelspin — and a trajectory that will take you off into the bushes if you don't do something about it.

The Cooper S is an expensive car if you want to maintain it properly. Apart from regular oil changes, which are essential for any sort of performance engine, the S demands a constant program of preventive maintenance to keep it in top trim.

Things like valves, which cost five times more than Mini valves, need attention every 16,000 km and in the same time the phosphor bronze valve guides and steel rocker shaft have normally worn too much to put back in. Similarly some Cooper Ss had an insatiable appetite for primary gear bushes, which, when they broke up, would distribute bronze grit throughout the engine and could ruin a motor.

A "best of everything" attitude would also see you replacing the Michelin XAS tyres on the front every 18,000 km (11,000 miles). The Michelins were by far the best to use, and would cling to the road long after any other radials had let go. But you paid plenty for the extra adhesion.

On the road from One Tree Hill to Gawler there are three sweeping left hand bends that dip down at the apex and then lead uphill and away. It is a favorite road for the young hoons of Gawler to go for a strop — especially early on a Sunday morning when most people are still having breakfast and all is still.

It was in that series of three corners you could really let it all hang out, screaming down into them with foot to the floor letting the understeer take care of the excessive speed build-up. Here, if you had the right tyres, rear wheels would lift and the door frame would twist so far out of shape it was amazing the lock didn't give.

Cold, irrefutable, performance figures showed that the Cooper S could not compare with a hot Holden EH manual, or slant-six Valiant in acceleration or top speed times. First gear was only good for 60 km/h (37 mph), second would pull 96 km/h (60 mph) and third would get you 137 km/h (85 mph). Flat out in top the often wildly inaccurate Smiths speedometer would show an optimistic 115 mph — but I knew that my car, which had been subjected to a cam grind and modded head, would only pull 6300 rpm in top or 162 km/h (101 mph).

But it felt fast, everything was

The Mini Cooper S was a natural for modification. This is one of the best developed Ss seen in Australia, an imported BMC works rally Cooper which Andrew Cowan drove in the 1970 Rally of the Hills.

happening around you, and through the Adelaide Hills or Blue Mountains the Cooper S more than held its own.

Now it's dead as a performance car. The introduction of the Bathurst Special Cortina 500 started its decline. The release of the XR Falcon GT put another nail in the coffin and when GMH introduced the Torana GTR it was all over for the Cooper S.

But don't dismiss the performing shoe-box entirely — there are those who still enjoy a Cooper S, and good examples of the marque are becoming showpieces.

The Cooper S is on the way to becoming a classic. It was the proof of the pudding that a good little 'un COULD beat a good big 'un. Ten years from now the Cooper S will be revered as THE cult car of the mid-sixties.

David Clement tells us that a good Cooper S can still be had for $1400. That's as low as they'll ever get, for, as the cars get harder to find, the price will rise. So if you're looking for a modern day classic, think of the Cooper S — the little horn car with the big, big heart.   *

Mini Cooper

We look back at the ultimate 'minibricks' and **Michael Bowler** tries a pair of Cooper S against the modern equivalent 1275GT; the three variations on a theme are shown opposite

WHEN the Mini arrived in August 1959 it asked to be tuned; it had all the roadholding but its 62·94 × 68·26mm, 848cc A-series only mustered a meagre 34bhp at 5500rpm. Specialists like Speedwell, Downton and Don Moore were quick to apply their knowledge developed from more conventional A-series designs such as A35, A40 and Morris Minor.

It took two years before BMC responded with the association of John Cooper to produce the first of the Mini-Coopers. That arrived in September 1961 with a 62·43 × 81·28mm configuration for 997cc and 55bhp at 6000rpm; it was a relatively long stroke but it had a much stronger crankshaft, new camshaft, bigger inlet valves, disc brakes and twin SUs.

The arrival of the competition inspired Cooper S in April 1963 with 1071cc seemed a curious size to choose, but it tied up with the 1100cc Formula Junior limits, even if it fell halfway between the 1000 and 1300cc saloon car classes. New blocks with centre cylinders moved in $\frac{1}{8}$-inch to allow a 70·64mm bore, coupled to a new nitrided crankshaft of 68·26mm, boosted the output to a healthy 68bhp at 5750rpm. Bigger discs with a brake servo looked after the stopping from higher speeds.

The Mini's popularity in saloon car racing prompted the next change in March 1964, when 970cc and 1275cc versions were added, but with 61·91mm and 81·33mm cranks and a taller block for the 1275S, giving 65bhp at 6500rpm and 75bhp at 5800rpm respectively.

Meanwhile the old Cooper had taken on the 998cc unit in January 1964, with its 64·58 × 76·20mm dimensions giving the same power output, but the car adopted the Hydrolastic suspension which was not really a success on such a short wheelbase. This was subsequently adopted for the Mk 3 Cooper S, by which time the suspension units had been modified, with blues and reds used to denote the competition versions. In that form the suspension wasn't too bad and stayed with the car till it dropped from the range in June 1971, the end of an era which had seen the little cars going ever faster in races and winning major rallies outright.

Its place was nominally taken by the 1275 GT with the same engine dimensions in less sophisticated materials and in totally emasculated tune, using a single carburet-

tor to give just 54bhp at 5250rpm – the same as the first Mini Cooper of 1961.

With so much recent talk of replacement Minis it seems appropriate to look back at the classic Cooper S versions and to make a comparison with the present day "hot" Mini. To do this we borrowed a 1275GT and journeyed east to see Philip Splett in Thorpe Bay. As an appreciator of the old Mini scene, he has acquired a pair of S's, a 1071 and a 1275. The 1275 is one of the very last produced, registered in early 1972, and is a quick one; the 1071 is virtually standard in specification, but is painted in the original Cooper Car Co col-

ours under which the works cars raced in their heyday.

I've always liked the 1275S, partly from driving on the road, but also from watching their racing antics with John Rhodes' technique particularly memorable – a well-timed chuck to get the car pointing in the exit direction as it starts the corner, thus reducing time-wasting understeer to zero.

Getting into this hot 1275 didn't immediately provoke the correct memories; it was certainly fast, the gears didn't whine like they used to, but it felt curiously unstable – a bit like a Renault Dauphine – until you really started to use it, when the handling was astonishingly neutral. In fact, it was a well sorted hot Min.

Although it is a Mk 3, this 1275 had been converted back to "dry" suspension, complete with all the correct parts to give negative camber at front and rear; this, combined with wide tyres on 6in Minilites, gave instant steering response and a lot of grip, with the result that there was very little power understeer and the back just followed round; equally there was little attitude change when lifting off the throttle. This hasn't affected the ride, which was never brilliant, but is quite acceptable for a quick little box.

The engine was certainly "hot"; a stage three head, with 1·48in inlets and 1·15in exhaust valves and a 12:1 compression, breathes through a pair of 1½in SUs with the assistance of a 288deg dwell steel cam prepared for the BMC engines that Ken Tyrrell used in the FJ Coopers. Lightened flywheel and fully balanced moving parts complete the picture.

Philip is still only the second owner of

 99

PHK 161M

Clad in Cooper Car Co colours, this is otherwise a virtually standard 1071S which goes as well as ever.

Mini Cooper

this projectile, which was acquired with a mere 4000 miles on the clock and hasn't been used that much since. It still is a tremendous fun car with masses of performance, although the cam robs the bottom end of much of the original torque, and a delightful short travel gear linkage. That I didn't feel instantly at home was due to its departures from standard, albeit period options, but it didn't take long to catch up and the special Corbeau seats were surprisingly comfortable.

Despite its looks the 1071 is virtually standard. In fact, it was built up from a 1974 Mk 3 car that had been used by a driving school. Its engine was a brand new one that had been stored till the car was built up in 1975; even now it only has 1700 miles on. The trim came from the other car which had been retrimmed to match its Corbeau seats.

This was much closer to memories of Cooper S with the power understeer more readily provoked and still with plenty of performance. It had the later gear linkage so the movement was longer than on the 1275, akin to that of the 1275GT. This also used dry suspension, but the ride, even on 70 per cent radials, was just firm and direct without being crashy; where the 1275 was set up to be driven round the corners, the

1071 could be chucked at them and traditional Mini handling enjoyed.

Although Philip Splett does a certain amount of his own work, general restoration and maintenance of these two cars had been entrusted to Gordon Dawkins of Carlow Engineering, Fulton Road, Manor Trading Estate, Thundersley, Essex, whose directors Tony Bunton and Gordon have been long known in the field of Mini racing over the past ten years or so. They have also restored such cars as Sprites and an Alvis TD21. If the standard of those two Minis is anything to go by, they know their stuff, bodily and mechanically.

Getting back into the 1275GT underlined the improvements that have been made to the car over the years. When it first arrived in 1969, the 1275GT was a big disappointment, particularly as the 1275S was still around. At that time the basic Mini cost £596, the 998 Cooper £710, the GT £834 and the 1275S £942. All used wet suspension except the new Clubman Estate. Problems with the original 1275GT were the use of the low 3·65 final drive against the latest Clubman 1000 which used the 1275S 3·44, so it was very fussy when cruising quickly. The new eyeball ventilation trunking passed very close to the radiator fan, so that transmitted more noise when the vents were open, and finally, wet suspension, with a lot of torque and a short wheelbase, was totally unsuitable — it upset the handling and created a lot of headlight direction variation from braking to acceleration. The 1275GT engine was a single carburettor version of that used for the Sprite from October 1966; this used a new block casting but kept the 1275S crank-size if not its mater-

ial specification. The camshaft still used the 230/252 deg dwell angles that started with the 948 Sprite and was used subsequently for all A-series derivatives, apart from the 997 Cooper which had 252 deg for inlet and exhaust. In that form the 1275GT developed 58bhp at 5250rpm against the twin-carb 998 Cooper's 55bhp at 5800rpm, but the benefit was obviously in torque, with 69lb.ft at 3900rpm against 57 at 3000.

In its favour the car's new nose gave a bigger car feeling, as well as providing better access under it, and crushability. Seating was more comfortable and the ride for normal motoring was better, but so too was that of the Clubman Estate than earlier dry Minis.

Fortunately all the failings of that early 1275GT have now been corrected. It is back to dry suspension, the fresh air duct has been re-routed and the final drive ratio is 3·44. The overall effect is of a very pleasant baby car, refined, comfortable enough and as agile as ever. Performance remains similar with a maximum bordering on 90mph, while fuel consumption over 750 miles of commuting and 70mph cruising was 35·1mpg. The comfortable cruising speed as far as noise is concerned is around 65mph — 4000rpm; in fact, one thinks wind noise is pretty low, until you try coasting from that speed and it is evident, but drowned by the engine. So, for motorway use, it could do with more sound deadening, but for any other motoring it is quite acceptable — an electric fan would probably make a worthwhile reduction.

The transmission is certainly quieter than it was, to the point of being no longer noticeable, and the gearchange is nice. Dunlop Denovos are standard fitting on the 1275GT and these give excellent grip in wet and dry, without feeling at all harsh as they had done in a back-to-back comparison on a Rover.

Although there is still some pitch and jerk in the ride it is noticeably more rounded than of yore, feeling pleasantly firm and sporting. Handling remains standard Mini with the grippier tyres increasing steering response and reducing power understeer. It is still a car that can flick through roundabouts with hardly a lurch and faster than most.

Inside, the old door pockets disappeared with the arrival of wind-up windows, but there is enough oddment space alongside the attractive three-dial instrument pod; the GT has a rev counter, which it doesn't need, whereas the faster Coopers were never so equipped. There can't be many cars on which you can adjust the left door mirror from the driver's seat, but the Mini is one of them, without being cramped for passengers. All-round visibility is excellent and the heated rear screen effective.

With comfortable seats that are rake adjustable, Mini driving is much more comfortable than it used to be and there are three positions for the seat anchoring bolts; I still had ¾in to go and was comfortable.

It was interesting to compare the new 1275GT with a pair of Cooper S's and see how well it came off; memory had it that the 1275S was a fantastic fun toy and that the 1275GT was a mistake. The passage of time and subsequent testing has closed the gap and the new car is a thoroughly modern Mini, very useable in town or country. But if anyone's selling a good 1275S, preferably with a lift-up tail-gate conversion, I might be interested in replacing the family hack with a Mini classic. ●

Outwardly standard this 1275S has a stage 3 engine conversion built up by Carlow Engineering although pancake air filters are the only obvious underbonnet changes. Inside, Corbeau seats give added comfort and location.

The stiff rear suspension lets the near neutral handling Longman Mini cock up a rear wheel.

Champion Mini

Richard Longman's Mini is heading for its second successive British Saloon title. What's it like to drive? By TONY DRON.

According to my copy of *The Eagle* dated December 5, 1959, the new Morris Mini-Minor and Austin Seven both use an 848cc, 34bhp engine, giving a top speed of 72mph and they can reach 60mph in third gear. According to Richard Longman, 20 years on, his Group 1½ Mini produces 105bhp (75 at the wheels) and can lap Silverstone at an average speed of 93mph, pulling 8000rpm in top with a good tow.

Nobody needs to be told that this is the car that currently leads the Tricentrol British Saloon Car Championship, and is a replica of the machine that Longman used to claim the 1978 title. Last year's car now resides in the private museum collection of Richard's sponsors, Patrick Motors. Richard was kind enough to transport his Mini to Snetterton for this AUTOSPORT track test, but at least it gave him the chance to put in a bit of testing for the possible Snetterton 24 hours Group 1½ race next year.

Richard reckons his car is worth about £6000, so I asked him just what exactly he does to the showroom product that turns it into a Group 1½ car. Naturally enough the first is to strip the machine down to the bare shell and scrape off the underseal to save a little weight. He does not bother with seam-welding the shell but just gets on with building the car up again. Just like the original 1959 Minis, the Patrick Motors Mini is fitted with Moulton rubber suspension, which in this case has been specially developed by Moulton for racing purposes. Adjustable Spax dampers are fitted and the car is set up with its 5J 12ins rims on positive camber at the back, negative at the front. Standard discs/drums are used with DS11 pads and VG95 linings, and the fully built-up racer weighs 830 Kilos.

The 105bhp engine, which Richard builds himself, produces its maximum power at about 7200rpm. Normally he revs it to 7600rpm in the specially homologated close ratio gears, but as we have seen he has proved it to be safe to 8000rpm in top with a tow. The excellent reliability of Richard's engines has not been achieved without cost. He spends hours on the rolling road, and he has deliberately destroyed two engines in testing this season alone by running them until they

Richard Longman offers a few tips as Tony Dron straps his helmet and prepares to go out.

cried enough. Then comes the painstaking business of stripping, examining, and taking notes — a system that has brought him to the conclusion that his cross-drilled, tuftrided crankshafts need to be replaced after every two or three races just to be on the safe side. Homologated options include a modified camshaft, twin 1½in SU carburettors, and a fabricated exhaust manifold which make up a punishing collection of goodies for a 1275GT crank to survive. But Richard keeps one step ahead of potential trouble by steady hard work.

He first raced in 1966, in a Mini at Llandow, and progressed to Escort Sports and Formula 3, taking pole for the GP meeting at Silverstone one year. Since than he has had a few one-off drives in a variety of machines, including that unfortunate Consul GT race at Castle Combe in 1973, when he broke his pelvis after a little encounter with Tony Pond. Since then he has returned to the Mini as the central theme of his racing efforts, eventually developing it until it was capable of beating Bernard Unett's rapid Avenger towards the end of 1977. His principal sponsor has stuck with him and nowadays you can buy the roadgoing equivalent of Richard's car in the form of a Patrick Motors Special Mini.

A season's racing means a budget of £20,000 or thereabouts, but if that seems expensive, rest assured that it's far less than you would need to tackle the bigger classes in the British Saloon Car Championship and stand a chance of winning.

Richard gave me no special instructions before letting me out of the Snetterton pits apart from: "It'll take you a few laps to get used to it, but just press on as long as you like until you're happy."

The first impression is that the car weaves on the straights. From the outside, this is not apparent, but from the inside it takes a couple of minutes to accept this sensation. Under braking, the car is ready to wag its tail even more, and it is not long before one is using this to set the car up for a corner. By setting up the back wheels with positive camber, Richard has deliberately built an amount of straightline instability into the car in an effort to avoid wasteful terminal power understeer. He has been completely successful.

As with any car, it is important to get the car travelling in an absolutely straight line before braking. It is possible to brake late and hard in the Longman Mini and, with a sensitive touch, it is not too hard to make a weave in the right direction coincide with the chosen turning point for a corner. Back on the power, the car then performs a neat trick, with the back end wanting to wash out and the front end wanting to move to understeer with the result that the two apparently undesirable tendencies have a little wrestling match and end up working in harmony. Once in this condition, the car is astonishingly stable and quick, and works well in slow corners as well as the fast ones. It just seems to rocket out of the awkward tight right-hander at the back of the 'new' Snetterton circuit, with the car seeming to float nicely while the front wheels put the power down with an effectiveness that certainly surprised me.

Near neutral

Like most front wheel drive racers, the Mini is set up with the rear suspension pretty stiff, and through the fast sweep of Coram I could feel it bouncing along behind as if it were almost solid. But the car stuck to its near neutral attitude admirably and I didn't have any mixed feelings about keeping my foot flat down in top.

While I was out on the circuit, there was a Jim Russell pupil having a go in one of the school Formula Ford cars. It was interesting to note that the Mini was much faster out of the tight corners, but by the end of the straight the FF predictably swept by the little saloon car. In fact the drag factor of the Clubman Mini is some 8 per cent worse than that of the original Mini shape, but unfortunately for Richard Longman it is, of course, the Clubman shape that has to be used.

Sticking to my usual rule for track tests, I pulled in when I began to enjoy myself so as to avoid the temptation of trying to set a personal best time. Even so, I got down into the 1min 22secs bracket, whatever that may mean. It's a bit quicker than I expected, anyway, but then I hadn't expected the car to be so good at putting its power on the road or to be so well balanced in the corners. After the recent Mallory Park fiasco, where the weather made nonsense of the current Tricentrol form, a further win for Richard has put him four points ahead of the field. I am pleased to say that my 10-lap thrash in the car seemed to have no ill effects, for he used the same engine at Mallory as for the track test. Time for another routine crank replacement, methinks.

It seems hard to think of the Mini as a 20-year-old veteran still plugging away on the circuits, but when you realise that in this form it is now capable of beating it's Group 2 counterparts of a few years back and still getting faster all the time, you have to accept that there's plenty of life in the old dog yet. Of course, the latest Dunlop tyres have made a big difference to lap times, even though with these unusual wheel sizes Richard does not have a wide range of compounds to choose from. Nevertheless, thanks to Mr Longman's hard work over the years, and his backing from Patrick Motor Sport, supported by ICI petrol and Duckhams Oils, it looks as if he's going to pull off the title of British Saloon Car Champion again. If you want to try to stop him, Richard Longman will be happy to sell you an identical car. Just write to him at Richard Longman & Co Ltd, 11-13 Purewell, Christchurch, Dorset, BH23 1EH! ∎